AF541069

ASTRAL

DICTIONARY OF PLANT SCIENCE

DICTIONARY OF PLANT SCIENCE

by

Dr. R.P. Chandola

2010
DAYA PUBLISHING HOUSE
Delhi - 110 035

ISBN 81-7035-658-X
ISBN 978-81-7035-658-5

Published by : **Daya Publishing House**
1123/74, Deva Ram Park
Tri Nagar, Delhi - 110 035
Phone: 27383999
Fax: (011) 23260116
e-mail : dayabooks@vsnl.com
website : www.dayabooks.com

Showroom : 4760-61/23, Ansari Road, Darya Ganj,
New Delhi - 110 002
Phone: 23245578, 23244987

Laser Typesetting : **Classic Computer Services**
Delhi - 110 035

Printed at : **Chawla Offset Printers**
Delhi - 110 052

PRINTED IN INDIA

Preface

Words in any language have always taken a major role in the advancement of human civilization. With the passing of time many words automatically added and enriched the various cultures towards provisions of a better, secured and longer life of all the living beings on this planet–the earth. Words have been used in the area of education which in turn has played a significant role towards improvement and upgradation of human life. In this regard scientific knowledge acquired through modern scientific research and education has influenced the human life tremendously vis-à-vis all other connected human activities.

Agricultural science including animal science as a particular way of life as adopted by human beings and also moved in the same direction as other sciences and also acts made good progress only primarily by means up many words with passing out of time. As far as scientific development for further improvement is concerned, it firstly evolved the scheme of making dictionaries to make it possible for easy comprehension of human behaviour, actions and reactions towards a civilized society. Such works also helped and emerged as one of the most important way in achieving an easy consultation for an achievement of rapid and significant knowledge of the different disciplines of agricultural science. In the present vastness of ocean of knowledge of any subject, it is difficult and cumbersome find the correct words for the different functions of agricultural science. The situation has become more tedious as the worlds pertain to the scientific and technical notations of various agricultural branches of knowledge have become numerous.

With the allure in view, six volumes relevant to the various subjects of agriculture *e.g.* Animal Science, Entomology (insect pest study), Crop Science (growing various agricultural field crops, agronomy), Plant Science (plant breeding and genetics), Plant Pathology (plant diseases), and Soil Science have been prepared with all scientific names (terms) used in agricultural education and research with comprehensive notes for various terms. Whenever possible some entries have been illustrated to make an easy understanding possible. Such a dictionary should also be used as a quick and easy reference to students and other concerned groups.

Dr. R.P. Chandola

Abelmoschus esculantus Family Malvaceae. Green fruits used as a vegetable.

Abrus precatorius Indian liquorice plant. Fam. Leguminoseae.

Absolute linkage During meiosis of the F_1, the two homologous chromosomes separate from each other and one of them goes to one pole and other to the opposite pole. Since the two genes are on the same chromosome, they always remain together in the same combination as that in the parent as they pass from one generation to another.

Absorption Roots and leaves are the main absorbing organs of plants. Roots absorb water and dissolved minerals or organic salts from the soil, while leaves take in gases–oxygen and carbon dioxide from the atmosphere.

Abutilon indicum Fam. Malvaceae. A common weed.

Acacia catechu Fam. Mimoseae. Yields a kind of tannin called catechu which is obtained after boiling chips of heart wood.

Acacia moniliformia Fam. Mimoseae. *Australina acacia*; yields gums.

Acalypha tricolor Bears floral spikes with pendulous axis which bears unisexual flowers only.

Accessory chromosome Supernumerary chromosome which is not necessary for the individual and not homologous to any of the normal chromosomes. This term was earlier used to denote a sex chromosome.

Acclimatization The adjustment of a species in the course of several generations to a changed environment.

Accumulated temperature *See* Crop Science.

Acentric Chromosome fragments without centromere.

Achene A small dry one seeded fruit developing from a single carpel but unlike the next one, the pericarp is free from the seed coat.

Acicular A leaf-shape; when the leaf is long, narrow and cylindrical, *i.e.* needle shaped as in pine, onion etc.

Acid lava A stiff viscous molten rock that flows slowly, has a high melting point (about 850°C) and congeals into thick tongues before having travelled far from the volcano vent.

Aconitum ferox Fam. Ranunculaceae. Also known as monk's hood or aconite. Is a medicinal plant.

***Acorus* sp.** The stem has a conjoint concentric bundle where the phloem lies in the centre and is surrounded by xylem and is the specialty of this monocotylednous plant.

Acquired characters Detailed studies have failed to show that acquired characters are inherited. Most biologists have therefore abandoned the theory of inheritance of acquired characters.

Acropetal succession Bearing of the vegetative parts of the plants in an order where the older and the longer parts are away from centre and young and shorter ones towards it.

Actinomorphic It is symmetry of a flower where it can be divided into two equal halves by any vertical section passing through its centre.

Active gene Suppose except for the difference in the two loci the two plants have the same genotype which is responsible for a given height. If there is difference in the plant height of any two plants in the progeny is due to the four duplicate (cumulative) incompletely dominant genes designated by capital letters, are called contributing or active genes.

Active transport Movement of ions or molecules of a chemical substance through the plasma or cell membrane from a solution of low concentration towards the solution of high concentration *i.e.* against electro-chemical gradient. It needs energy and comparable with passive transport (osmosis, diffusion) which require no energy.

Acuminate It is the shape of the apex of a leaf, when it is drawn out in a long slender tail. Also called caudate.

Adaptations According to the Law of Evolution as postulated by Lamark, Jean Baptaste, change in habits evolves extra use of some organs and less use of the others. Consequently there occurs some change in the organs of the organism. These changes were named adaptations.

Addition line An addition line has one pair of chromosome from another variety or species in addition to the normal somatic chromosome complement of the species.

Additive gene Non-allelomorphic genes affecting the same character and enhancing each other's effect.

Additive ratio In barley plant, light purple colour of the grain is evidently due to the presence of a dominant gene P_1 or another dominant gene P_2. The two non-allelic dominant genes P_1 and P_2 posses an additive effect and the colour of the grain is dark purple when the genes P_1 and P_2 are present together. When both dominant genes are absent, the colour of the grain is white. The genotype ratio of F_2 in such case is the 9 dark: 6 light 1 white.

Adenine A base material lying on the flat planes of the spiral of DNA molecule like steps in a spiral staircase. These are four other base materials also situated in the same patter *i.e.* guanine, thymine and cytosine ($C_5H_5N_5$). A purine base, 6-aminopurine, occurring in the nucleic acid.

Adenosine triphosphate (ATP) An energy rich compound participating in energy restoring and energy using reactions in the cell.

Adhesion and cohesion The term adhesion (adnate or adherent) is used to designate the union of members of the different whorls *e.g.* petals with stamens; or stamens with carpels; and cohesion (connate and coherent) to designate the union of members of the same whorl *e.g.* stamens with each other, and carpels with each other.

Adiantum In this genus and also in the genus *Ptreis* both belonging to tile *Pteridophytya,* the rhizomes are branched, creeping strangling type and bear ramenta.

Adiantum caudatum A kind of walking fern, propagate vegetatively by their leaf tips.

Adina cordifolia Fam. Rubiaceae. An ornamental plant.

Adnate stamen When the filament runs the whole length of the anther from the base to the apex.

Adnate stipules These are the two lateral stipules that grow along the petiole up to certain height, adhering to and making it somewhat winged in appearance.

ADP Adenosine diphosphate.

Adventitious roots Sometime the tap root developed from the radical stops growing and dies out after a short while and more roots are developed from other parts of the plant. Such roots which develop from any other part of the plant from any other part of the plant except the radical are known as adventitious roots.

Aegilops The hexaploid wheat is believed to have originated between a tetraploid wheat (genomes-AABB) and Aegilops species (2n-14; genomes DD).

Aegle marmelos Fam. Rutaceae. Wood apple.

Aerenchyma Star like parenchyma, with radiating arms, leaving a lot of air cavities. If the parenchyma tissues has very large air cavities separated by partitions of cells, the tissue is known as aerenchyma.

Aerobic respiration In all the plants and in the living cells of a plant, respiration in the presence of air consists especially in the oxidation of substances like carbohydrates, fats and some other organic materials. Generally the oxidation is complete and the end products are carbon dioxide and water. If the substrate is a hexose sugar, the reaction of the process will be: $C_6H_{12}O_6 + 6O_2 = 6CO_2 + 6H_2O + \text{energy (ATP)}$.

Aestivation The mode of arrangements of the sepals or of the petals, more particularly the latter, in a floral bud with respect to the members of the same whorl.

Agaricus The common mushroom, is a fleshy saprophytic fungus. It has edible and poisonous forms.

Agave This *Aloe* sp. Has sunken stoma in its leaf.

Ageratum conzoides Fam. Compositeae. A common weed, plants have medicinal value.

Aggregate fruits It is a collection of simple fruits (or fruitlets) developing from an apocarpous pistil (free carpels of a flower).

Agouti In wild mice, the colour of the coat is greyish brown. The colour blends with the environment in which they live and affords protection against enemies and is called agouti. When black mice are crossed with albinos, the progeny is found to be agouti. When these agouties are inbred, agouti, black and white appear in a ratio of 9 agouti: 3 black : 4 white.

Agronomy The science of managing the land. It includes the theoretical background and practical aspects of growing crops, maintain the soil's fertility and the rearing and caring of farm animals.

Air canal In the assimilatory region of *Riccia* (Bryophyta), there exist vertically running narrow slit called air canal or air spaces.

Akinetes It is a vegetative cell; the wall of which get thickened by additional layers of cellulose and pectose as in the case of *Spirogyra farlowii.*

Albino A plant lacking chlorophyll or an animal lacking pigmentation.

Albizzia lebbek Fam. Memoseae. A useful timber plant.

Albugo Synonym–*Cystopus.* Derives its name due to the formation of white patches on the surface of infected host. The fungus consists of a number of biological races or *formae speciales*, each of which is restricted to an individual species. *See* Plant Pathology.

***Albugo* spp.** Causes common downy mildew diseased. It grows as a parasite on many plants of the mustard family. It causes a disease called 'white rust'.

Albuminous The endosperm is the food storage tissue and when it is present the seed is said to be endospermic or albuminous. In many seeds, however, all the food stored up in the endosperm at its early stage is used up by the developing embryo. The endosperm is thus absent in such seeds, and they are then said to be non endospermic or exalbuminous.

Alburnum The out region of the secondary wood which is of lighter colour is known as the sap wood or alburnum.

Alcaptonuria An inherited metabolic disorder in which there is an excretion of excessive amounts of homogenetisic acid (alcapton) in urine.

***Aldrovanda* spp.** It is a rootless, free floating plant with whorls of leaves. The mechanism of the prey is the same as that of Dionaea, but instead of only six sensitive hairs there are a number of them here on either of the midrib.

Aleurone grain Each aleurone grain is a solid, ovate or rounded body and encloses in it a large crystal like body, known as crystalloid, and a small sounded mineral body, called the globoid. It is a common form of insoluble or sparingly soluble protein, abundantly found in the endosperm of seed.

Alga These are the thallophytes containing chlorophyll and sometimes other pigments.

Algal zone In the coralloid root of *Anabaena cycadeae* (Gymnosperms), the cells of root zone contain blue green algae like radically elongated cells in about the middle of the cortex. This is called the algal zone.

Alien addition line It has one pair of chromosome from a related wild species in addition to the normal somatic chromosome (2n) of the species.

***Alisma* spp.** Known as water plantain.

Allete One of a pair, of geries, of forms of a gene situated at the same locus of homologous chromosome. Also called allelomorph.

Allium cepa Fam. Liliaceae. Onion, an important vegetable.

Allium sativum Fam. Liliaceae. Garlic, a common vegetable.

Allogamous Reproducing by cross pollination.

Allopolyploid An organism with more than two sets in its body cells, derived from two or more species by hybridization. This term is often abbreviated as alloploid.

Allosteric effect Reversible interaction of a small molecule with a protein molecule causing a change in the shape of a protein and consequent alteration of the interaction of that protein with a third molecule.

Allozymes Alternative forms of an enzyme code by different allele at a single locus.

***Allysum* sp.** An ornamental plant. Aliason.

Almond Fam. Rosaceae. *P. amygdalus.* A nutritive fruit grown in temperate climate.

Alternation of gernation The life history of any of the higher cryptogams *e.g.* liverworts, mosses and ferns, is complete in two stages or generations, alternating with each other. These two generations differ not only in their morphological characters but also in their modes of reproduction.

***Althea* sp.** Hollyhock, used as an ornamental plant. Fam. Malvaceae.

Amanita Certain species of amenitis which resemble agaricus, are extremely poisonous, however, they are usually distinguished from the latter by their possession of a cup-like structure at the base, which is wanting in agaricus.

***Amaranth* sp.** Fam. Amaranthaceae, is an important green leafy vegetable.

Amaryllidaceae This family has a same general characters as the Liliaceae, but differs from the latter in its ovary being inferior.

Ameboid movement It is the creeping movement of naked masses of protoplasm *i.e.* not enclosed by cell wall.

Ameiosis Non-reduction of chromosome numbers due to suppression of one of the mitotic divisions.

Amino acids A class of chemical compounds containing an amino (NH_2) group and a carboxyl (COOH) group, plus a side chain; the basic constitutional unit of proteins.

Amitosis Direct nuclear divisions, without the separation of daughter chromosomes.

Amneiocentesis A procedure for obtaining amniotic fluid from a pregnant woman for diagnosis of some disease resulting from chromosomal irregularities.

Amphicribral In a concentric vascular bundle, xylem forms a central core surrounded by phloem or vice versa. In the first case, the bundle is said to be amphicribral.

Amphidiploid A polyploid originated by hybridization between two species or genera and whose somatic chromosome complement is made up of the entire somatic complements of the two species. Also called amphiploid.

Amphimixis To bring together the elements from two gametes in fertilization. The union of egg and sperm in sexual reproduction.

Amphitrichous In bacterial form when a tuft of flagella is attached at either end.

Amphitropous When the ovule is placed transversely at a right angle to its stalk or funicle.

Amphivasal in a vascular bundle, phloem is surrounded by xylem.

Anaerobic phase This is the initial phase of respiration in plants in which a simple carbohydrate like glucose undergoes in complete oxidation leading to pyruvic acid. The process is called glycosis, and it takes place in the absence of oxygen (strictly speaking, oxygen is not at all required for this process). Glucose is first phosphorylated *i.e.* a phosphate group added to it by ATP (an active energy rich phosphate compound). Glucose phosphate is formed. Reaction follow in quick successful as glucose phosphate-fructose phosphate-phospoglyceraldehyde-phosphoglyceric acid-phosphopyruvic acid (with the removal of phosphate).

Anaerobic respiration. When the plants are deprived of oxygen, they respire anaerobically; although there are some plants which do not require oxygen at all. They wholly depend on anaerobic respiration. Here the substrate is partially broken down and the end product as a rule being carbon dioxide and ethyl alcohol. In case a hexose forms a substrate, the whole process goes like this: $C_6H_{12}O_6 = 2C_2H_3OH + 2CO_2$ + energy. This type of respiration takes place in some bacteria like denitrifying bacteria and methane bacteria etc. These bacteria are obligate anaerobes

and bring about reductive conversion of carbon dioxide to methane. Hydrogen formed as a result of glycolysis is used in this reduction: $4H_2O + CO_2 = CH_4 + 2H_2O$ + Energy
Methane

Analogy Study of organs from the point of their identical structures and functions.

Ananas comosus Pineapple. An important fruit. Fam. Anonaceae.

Anaphase The stage in nuclear division during which the members of pairs of chromatids or chromosomes separate and move to different poles of cell spindle.

Anatomy Study of gross internal structures of various plants or animals as seen in a microscopic section or without. The study under a microscope is known as histology.

Anatropous When the ovule bends back alongside the funicle, so that the micropyle lies close to the hilum, the microphyle and chalaza, but not the funicle, lie on the same straight line; this is the commonest form of ovule.

Androcytes In *Riccia* (Bryphyta), the last division of the androgonal cells results in the formation of androcytes or sperm mother cells and each one of the androcytes gets metamorphosed into a male gamete, the antherozoid or spermatozoid.

Androecium Is the male whorl of a flower. It consists of three parts (1) stamens made up of filament anther and connective.

Androphore When the internode between the corolla and androecium is elongated considerably, it is known as androphore.

***Anemone* sp.** Fam. Ranunculaceae. Ornamental plant known as wind flower.

Androgenesis The phenomena of parthenoganesis by male.

Anemophily Pollination is brought about by wind.

Anethum graveolens Fam. Umbellifereae. Sowa plant, used as a leafy green vegetable.

Aneuploid An individual whose somatic nuclei do not contain an exact multiple of the monoploid (or basic) number of chromosomes.

Angiosperms One of the two subdivisions of Phanerogams.

Angrostrom 0.0000001 mm.

Angular divergence In the phyllotaxy, in such cases the genetic spiral makes one complete turn, subtending an angle of 360° in the centre of the circle and it involves two leaves; so the angular divergence, *i.e.* the angular distance between any two consecutive leaves is of 360 or 180°.

Animal pole That portion of egg which has active crystoplasm and zygote nucleus.

Anisogamy The fusion of unequal sized gametes. Form like C *Chlamydomonus braunii* (algae) the male cell reduces four large macrogametes and the male cell 8 small microgametes. The female gametes are larger and less active while the male gametes are smaller and more active.

Annona sqamosa Fam. Amonaceae. Custard apple. A fruit tree.

Annual rings In spring season, the cambium becomes more active and forms a greater number of vessels with wider cavities (large pitted vessels); while in winter it becomes less active and elements of lesser dimensions (narrow pitted vessels, trachieds and wood fibres). The wood thus formed in the spring is called the spring wood or early wood, and that formed in winter is called autumn wood or late wood. These two kinds of woods appear together and are called the annual ring or growth ring and in a transverse section is seen as concentric rings. In plants growing in such parts of the world where the climatic conditions are not favourable for growth throughout the year, the activity of the cambium also varies in different parts of the year. In temperature climate, the cambium is most active in spring and the secondary wood formed during this period is extensive and consists of large vessels known spring wood. Thus according to seasons, various rings of annual rings are formed in two parts–the spring wood and the autumn wood.

Annuals Those herbs that live for a few months or at most for a year.

Annular This is pertaining to the secondary thickening of the cell wall. When the deposit of the lignin is in the form of rings, it is called annular.

Annulate lamellae The pores and lamellae (*i.e.* the pore complex) of nuclear envelop of most eukaryotes and endoplasmic reticulum of invertebrates and ovoçytes and spermatocytes of vertebrates.

Annulus In Bryophyta (*Fumaria* sp.), the inner layers of operculum or lid are thin layered cells and form the major portion of the operculum. In between the operculum and body are present perestome and annulus.

Anthecephalus indicus Fam. Rubiaceae. *kadam* plant. Medicinal uses.

Anther *See* androecium.

Antheridiophores In the thallus of Marchantia, on the upper surface it bears a number of cup-like outgrowth, known as gemma-cups, on the mid rib the male plant bears some special erect male reproductive branches, the antheridiophores, each with a more or less circular disc or receptacle with radiating rays or arms.

Antheridium In the sexual reproduction of algae, fusion of two differentiated gametes–the male one called antherozoid, is very minute, ciliate, and mitile, while the female one is called the' egg cell or oosphere, is large and passive. The antherozoids are borne in pairs in a cell called antheridium.

Anther lobes The anther has two halves known as anther lobes.

Antherozoids In higher algae, fungi, mosses, ferns and allied plants, male gametes are very minute, motile, ciliate and active and are known as antherozoids or spermetazoids.

Anthesis The process of dehiscence of anthers; the act of flowering.

Anthocynin Most colours of flowers such as violet, purple, blue, brown and often red are due to presence of this pigment, which remain dissolved in the cell sap.

Antibiotics These are toxic chemical substances, possibly enzymes, secreted by certain soil bacteria or soil fungi, which have a destructive effect on particular disease germs invading the body and causing infectious diseases often of a serious nature *e.g.* pneumonia, typhoid, diphtheira, tuberculosis, cholera, erysipilas etc. Antibiotics are the drugs which often are used to shoot down these microorganisms and save human being and animals from such ailments.

Antigonon corculum The sadwitch Island climber plant.

Antipodals The three nuclei at the chalazal end of the embryo sac are called antipodals.

Antipodals cells At the opposite end of the embryo sac there is a group of cells, three in number known as antipodal cells.

Antirrhinum Snapdragon. This plant is an example of incomplete dominance. A plant with crimson petals crossed with one with white petals gives first generation with flowers that are all pink and the segregation in the F_2 as 1 red: 1 pink: 1 white.

Aplanospores In algae under unfavourable conditions, the zoospore instead of being liberated may round off to form thick walled aplanospores.

Apocarpous In a compound or polycarpellary condition, when the carpels are free, called apocarpous condition.

Apocynaceae Family of flowering plants.

Apomeiosis Meiosis that is either suppressed or imperfect.

Apomixis The occurrence of the external form of sexual reproduction without the occurrence of meiosis and fertilization.

Apophysis It is the basal solid portion of a capsular fruit having chloroplast in many cells and stomata in the epidermis. It is the basal portion of the capsule of *Funaria* (Bryophyta). Also in the ovuliferous scale which is borne on the upper surface of this bract scale in its axial. It is wedge shaped in outline and its distal, broader, sterile portion is called apophysis. This pertains to coniferous plants.

Apple *Pyrus malus.* Fam. Rosaseae. An important temperate fruit.

Applied climatology The use of the prediction (forecast), science of climatology in the predictions both long and short term. This is of particular importance in agricultural planning.

Apposition Growth of the cell wall in thickness is mainly due to deposition of definite thin plates or layers by the protoplasm, one after another, one the inner surface of the original wall. This method is known as growth by apposition. During the growth of any cell wall, cellulose is added in the form of new layers on the inner side of original wall; thus increasing the growth of cell wall takes place only in this way.

Arachis hypogea Family Papilionaceae. Groundnut. A very important oilseed crop of tropical and subtropical climates.

Archegoniophore The female plant in Marcantia, bears special female branches, called archegoniophores, with a star shaped disc, or receptacle.

Archegonium It is a flask shaped body consisting of a swollen basal portion called ventre and a narrow tubular portion, the neck in Marchantia.

Arc indicator Also called lever auxanometer. Growth is very slow in plants, but it can be accurately measured with the help of this instrument.

Archicarp In *Aspergillus sp.* (Fungus) in sexual reproduction, the male and female sex organs are developed near each other on the same or separate hyphae of the same mycelium. Hence it is homothallic. The male sex organ is called antheridium whereas the female sex organ is called archicarp.

Archegonial chamber In *Cycas*, the vegetative tissue surrounding the archegonia grows very rapidly to form a cavity called archegonial chamber.

Archegonium This structure like the antheridia in *Riccia* (Bryophyta) are also borne in the mid-dorsal groove and each one is enclosed in a cavity formed by the vegetative tissue around it.

Area Any expanse forming a region, district or locality.

Areca catechu Fam. Palmaceae. Beetle nut palm.

Aril In some seeds, a fleshy outgrowth arises from the mycropyle or the funicle which completely surrounds the seed and is called the aril as in Litchi.

***Arisaema* sp.** The snake or cobra plant, the spathe is greenish purple in colour and it expands over the spadix like a hood of a cobra. This acto of imitating the appearance, colour or any particular feature of another plant or animal is called mimicry.

Aristolochia gigas A censor mechanism of seed dispersal takes place and accordingly the fruit dehisces and when it is disturbed by wind, the seeds are thrown out.

Armature This is a condition in plants when their body may take form of arms or defensive weapons for self defence against the attack of herbivorous animals. These may be in the form of thorns, spines or prickles.

Artichoke (Jerusalam) *Helianthus tuberosus* has adventitious roots absent from the tuber.

Artificial epiphytotics In breeding for resistance, exposure to the disease is necessary to distinguish between the resistance and susceptible plants. Since a natural disease epidemics may not occur in the field every year, it would be necessary to establish disease epiphytotics by artificial means either in the field or in the greenhouse.

Artificial mutations Those mutations or genetic changes which have been evolved by use of artificial mutagenic agents such as X-radiation, Gamma-radiation and chemical mutagens.

Artimecia nilgiricus The Indian wormwood, used as a medicinal plant.

Artificial system of classification This type of classification is based on the artificial characters such as size and form, and plants were divided as herbs, shrubs and trees etc.

Artocarpus heterophyllus Fam. Moraceae. Jack fruit. An important vegetable or edible fruit.

***Arundo* sp.** The giant reed. Fam. Gramineae.

Asafoetida Fam. Umbelliferae. *Ferula asafoetida,* has medicinal value and is used as a condiment.

Ascent of sap The water absorbed from the soil by the root hairs slowly moves up through the plant body to the leaves and the growing region of the stem and the branches usually at the rate 1.2 m per hour. This activity is dependent on a few factors (1) path of movement of sap (2) root pressure (3) Cohesion forces (4) inhibition (5) activity of living cells (6) Pulsation of the internal layer of the cortex abutting upon the endodermis.

Ascogonium It is a long straight and somewhat club shaped body in Penicillium. At first it is uninucleate but later it becomes multinucleate by repeated divisions of the nuclei. A slender hypha (antherifodial) arises from the separate vegetative phypha and grows twining round the ascogonium, upto a certain height. Its swollen terminal cell is the anthredium.

Ascospore These spores are formed in the asci of the Saccharomycetes (fungi). The wall of the ascus breaks and the ascospores are set free. Each ascospore enlarges and becomes a somatic cell. It divides by a transverse wall into two daughter cells.

Asexual reproduction Reproduction that does not involve the union of gametes.

Aspergillus *See* Plant Pathology.

Assimilation It is the absorption of the simplest products of digestion and foodstuffs by the protoplasm into own body and conversion of these products into the similar complex constituents of the protoplasm.

Assimilatory region In Bryophyta, structure of the thallus is very much elaborated as compared to the external features. Vertical transverse section through the thallus shows that internally it is divided into two distinct zones or regions–a lower or ventral storage region and an upper or dorsal photosynthetic or assimilatory region.

Asynapsis Failure of pairing of homologous chromosomes during meiosis.

Atavism It shows the reversal of an organism to the ancestral (not parental) form. It is best illustrated by some citrus leaves which become trifoliate by the development of two leaves at the base of the lamina.

ATP Adnosine Triphosphate.

Atropa belladonna Fam. Solanaceae. Deadly nightshade. A medicinal plant.

Autocarp A fruit formed as a result of self fertilization.

Autocology It is the study of individual species in relation with the environment.

Autoclave To kill bacteria and spores of fungi, an apparatus is used in which use of dry heat or steam under high pressure is created to kill these organisms. This process is called sterilization.

Autogamous Reproduction by self fertilization.

Autogamous variations Heritable variations.

Autogamy Same as self pollination.

Autogenous chimeras These originate in somatic cells as a result of some natwal process by which diverse cells arise and give rise in turn to diverse tissues.

Autogenous variations Heritable variations may be caused either by recombination of genes or by mutations. They are called autogenous variations because they are due to internal causes *viz.* by changes in the genotype.

Autonomic It is a movement of locomotion which is spontaneous.

Autophagy The digestion of cell organalles by its own lysosome.

Autophyte All green plants prepare their own organic food, particularly carbohydrates, mostly in their leaves, from the raw of inorganic materials absorbed from the soil and the air, and nourish themselves with their food. Such plants are called autophytes.

Autopolyploid Is a polyploid arising through multiplication of a single genome. In other words, it is a polyploid of single species and all its sets of chromosomes are identical or very closely similar to each other. If it refers to a basic set of chromosome (*i.e.* genome), an autotetraploid may be denoted as AAAA.

Autopolyploid sterility Autopolyploids are usually highly sterile because the behaviour of chromosomes during meiosis in autopolyploids is peculiar due to the fact that they posses more than two homologous chromosomes.

Autoradiograph A photograph prepared by labelling a substance (DNA) with a radioactive material (tritiated thymidine) and allowing the image produced by decay radiation to develop on a film over a period of time.

Autosome A phenomenon in which sex specific, nongenital differential characters are displayed.

Autotriploid The genome of this is of the type AAA *i.e.* the basic genome is present in triplicate.

Autotroph An organism that obtains its energy directly from inorganic sources such as sunlight or a rock surface, and from these synthesises organic materials. For example, green plants are autotrophic. Thus autotrophs can be said as primary producers belonging to the first trophic level and thus supply energy to other life forms.

Autotrophic bacteria Such bacteria can synthesise their own food from an inorganic medium containing water, carbon dioxide and inorganic salts.

Autotrophs This kind of plants can derive their energy either from the sun or some energy yielding inorganic chemical reactions. These plants can synthesise all the organic compounds like water, carbon dioxide, nitrates and other trace elements in the presence of sunlight.

Autumn crocus Name of the plant *Colchicum autumnale* of Fam. Liliaceae which produces Colchicine, an alkaloid, reported to be very effective in doubling the chromosome numbers in plants.

Autumn wood *See* annual ring.

Auxanometer *See* arc indicator.

Auxins These are group of substances produced by the growing apices of stems and roots. They migrate from the apex to the zone of elongation in plants.

Auxotroph An individual unable to carry on some particular synthesis, hence requiring supplementing of minimal medium by some growth factor, a mutant organism (bacteria) that will not grow on minimal medium but requires the addition of some growth factor.

Averrhoa carambola Kamrak, Family Apocynaceae. An important fruit plant.

Axile A placentation, the ovary is two to many chambered usually as many as the number of carpels and the placentae bearing ovules develop from the central axis corresponding to the confluent margins of carpels and hence the name axile.

Azadirachta indica Margosa. Neem is an important tree used for medicine and timber.

Azotobacter Aerobic bacteria which fix atmospheric nitrogen in the soil.

Azygospore In algae (Spirogyra), in some cases of gametes fail to fuse. Such gametes behave like zygospores. They withdraw their cilia, round off and secrete a thick wall and perennate through the unfavourable period. Such spores are called azygospores.

Bacillus Cells of these bacteria are rod shaped and elongated. Commonly the daughter cells of bacilli remain joined end to end forming a filament of thread called *streptobacillus.*

Back cross A cross between a hybrid and either of the parents.

Back mutation Mutations are sometimes reversible. An allele that arose through mutation of a gene is likely to mutate back to its original form.

Bacteria By his newly constructed microscope Anton Van Leeuwenhoek in the late sixteenth century, he examined a drop of water under it and found a variety of living objects wriggling and darting across the field of his microscope. The word bacterium derived from a Greek word meaning "little rods" (or stick) was coined by a German Prof. F.J. Cohn in the middle of 19th century.

Bacteriochlorophill Some bacteria have a purple photosynthetic pigment present in them. This is related to (not identical) the chlorophyll of higher plants but differs from it in having the maximum power of absorption of radiant energy from the invisible portion of light *i.e.* from the infrared region (wave length 800–890 mu). Such bacteria are capable of synthesising their own food from the inorganic raw material. These are also termed as photosynthetic autotrophs.

Bacteriology Also called microbiology. It is the study of microorganisms like bacteria.

Bacteriophage The viruses attaching the bacteria cells.

Balanced translocations Positional changes of one or more chromosome segments without alteration of the normal diploid or haploid complement of genetic material.

Balance theory of sex determination Which sex actually develops is decided by the balance *i.e.* by the preponderance of the female-determining or of the male determining genes. The sex chromosomes are merely vehicles of genes which help in tilting the balance in one direction or another.

Bamboo *Bambusa sp.* Fam. Gramineae. Economically important plant. A plant group of large often large grasses *(bambuseae)*, native to tropical and subtropical areas *e.g.* monsoonal Asia. The strong woody hollow stems of bamboo may attain heights of 30 m in a few months growth. They are used as a common and cheap construction material for housing, boats and bridges. Young shoots are also eaten as vegetables. Bamboos thrive in deep fertile soils but each species fails periodically after a long delayed flowering and seeding that exhausts the plants energy.

Banana *Musa paradisiaca.* Fam. Museae. There are several varieties. The plant is propagated by suckers. The ripe fruit contains about 20 per cent of sugar but no starch and about 4.7 per cent protein.

Barassus flabellifer Like some other cultivated species like ganja plant (*Cannabis sativa*), and date palm dioecious condition is found in this plant, the plamyrah palm.

Bark All the dead tissues lying outside the active cork cambium constitute the bark of the plant. It therefore includes the epidermis, lenticels and cork and sometimes also hypodermis and a portion of the cortex.

Barley Fam. Gramineae. *Hordeum vulgare.* A useful plant, used as staple food and straw used as fodder for the cattle etc. *See* Crop Science.

Basic number The number of chromosomes, or assumed to have occurred in the gametes of diploid ancestor of a polyploid. Represented by x. The number of chromosomes constituting a genome. It is the number of chromosomes found in the gamete of a true diploid.

Basil Fam. Labiateae. Sacred basil, *Ocimum sanctum*.

Basin cultivation The growing of crops in small hollows or basins after formed by raising low ridges of soil around a plot of land. This type of cultivation is used by subsistence farmers in areas of tropics where rainfall is high and soil erosion common. The small hollows check rain wash and collect water for the growing crops.

Basin irrigation A method of providing water for cultivation that involves the flooding of basin like hollows and surrounded by earth banks. The hollows vary considerably in size.

Basipetal It is the order of development of conidia that shows that the oldest condiurp at the apex and the youngest near the base as in *Albugo.*

Bast The secondary phloem consists of sieve-tubes, component cells and phloem parenchyma and often also some bands or patches of bast fibres.

Bast fibres Also called phloem fibres. Sclerenchyma fibres present in the phloem are termed as bast fibres.

Bath sponge Fam. Cucurbitaceae. *Luofa* or *luffa cylindrica.* An important vegetable.

BC_1, BC_2 Symbols used to designate the first and second backcross generations.

Beaded root Also called moniliform root. In such roots there are some swellings in the root at frequent intervals.

Beans A big group of vegetable pods/legumes. Cluster beans, Soya bean, French beans etc.

Beet Beet-root is the source of sugar in some cold countries. Russia is the biggest producer of sugar beet. *Beta vulgaris.*

Benincasa hispida Fam. Cucurbitaceae. Ash or wax gourd.

Bentham and hooker system This system of classification of plants is also called the natural system. According to this system, the dicotyledons have been divided into Polypetaleae, gamopetaleae and monochlamydeae. The monocotyledons are divided into seven series.

Berry Also called Bacca. A fleshy, superior (sometimes inferior), usually many seeded fruit, developing commonly from a syncarpous pistil (rarely from a single carpel), with axile or parietal placentation.

Beta rays Radioisotopes of phosperous-P^{32} and sulphur-S^{32} can be used as a mutagenic agent. These are administered in such a way that they move in solution through the transpiration stream to actively dividing meristems.

Beta vulgaris Sugar beet triploids have been found to have not only larger roots but also higher yields of roots and sugar per unit area than diploids.

Betel *Piper betle.*

Betel nut It is used for chewing with betel leaf. It is borne on betel palms. Fam. Palmaceae.

Beverages They are agreeable liquors meant for drinking. Tea, coffee and cocoa are such common beverages.

Bicollateral This is a condition of a vascular bundle where both phloem and cambium occur twice, once on the outer side or xylem and then again on its inner side.

Bicollateral vascular bundles In certain cases as in *Cacurbits'* stems in each vascular bundle, there are two patches of phloem, one internal and other external to the xylem and two cambial strips– inner and outer between xylem and phloem. Such bundles are known as bicollateral.

Biennials These plants attain their full vegetative growth in the first year and produce flowers and fruits only in the second year after which they die off.

Bifacial leaf In some leaves, internally there is difference in the tissues below the upper and lower epidermis. These are known as bifacial leaves.

Bilabiate This is a zygomorphic gamopetalous corolla in which it is divided into two portions or lips, the upper and lower, with the mouth gaping wide open.

Binary fission The nuclear matter of the bacterium divides into two and is then followed by formation of a constriction across the long axes of the cell.

Bind weed (water hayeinth) Fam. Convolvulaceae. *Ipomoea reptans.*

Binomial expansion Exponential multiplication of an expression consisting of two terms connected by a plus (+) or minus (-) sings such as $(a+b)^n$.

Binomial nomenclature All the members of a species are related in the sense that they have descended from the same ancestors. Hence all the plants making up one species are alike in fundamental vegetative and reproductive characters. The most important character which can be used with some certainly for deciding that a given plant belongs to a species is that they can interbreed and produce fertile offspring. A genus is a group of similar and related species having a characteristic inflorescence, are included in the same genus. In binomial system of nomenclature, each specimen is given two names–the name of the species and the genus to which it belongs.

Biochemistry Some of the most recent and very important branches of biology are biophysics, biochemistry, biometrics and radiology.

Biogenesis The production of living things from living things.

Biogeography It deals with the distribution of the living organisms on the earth and various factors responsible for its distribution.

Biological race A strain which though morphologically indistinguishable, yet differs from the normal in its pathogenicity.

Bioluminiscence In both the types of respiration–aerobic and anaerobic, heat is produced which is sometime so enormous that stacks of grass or hay may catch fire. At times, this heat energy is converted into light energy and emits the phosphorescent glow. This phenomenon is technically known as bioluminescence and is responsible for the 'ghost light' of the forests.

Biometry Application of statistical methods to the study of biological problems. The branch of science which deals with the application of statistical methods to biological investigations.

Biophytum sensitivum The wood sorrel.

Biotic factors These includes the soil bacteria, algae, protozoa, earthworms, and burrowing animals which alter the soil often making it fertile.

Biotype Distinct physiological race or strain within morphological species having identical genetic constitution. A group of individuals all of one genotype. A biotype may be homozygous or heterozygous.

Bipinnate When the compound leaf is twice pinnate *i.e.* the midrib produces secondary axes which bear the leaflets.

Birch *Betula bhuriiapatra.* Fam. Batulaceae. An important timber tree.

Bisexual When both stamens and carpels are present, the flower is said to be bisexual or hermaphrodite.

Bitter sweet Fam. Solanaceae. *Solanum dulcamara.*

Bivalent Two homologous chromosomes held together by chiasmata in the first meiotic division.

Black cotton soil *See* regur, Soil Science.

Black mould It is the common name of *Mucor. See* Plant Pathology.

Bladderwort *Utricularia* sp. Mostly floating or slightly submerged plants, rootless, aquatic herbs, a carnivorous plant.

Bleeding Sometime it so happens that certain plants when cut, pruned, tapped or otherwise wounded, show a flow of sap from the cut ends or surfaces, often with considerable force. This phenomenon is known as bleeding.

Blending inheritance Inheritance in which clearly defined segregation is lacking in the F_2. This is generally due to the presence of multiple factors, independently inherited.

Blepheroplasts In *Chlamydomonus,* the flagella of equal size come out from the interior pointed end. These are fine hair like processes projecting either through a single aperture in the wall or through two separate canals. There is a single basal granule called blepheroplast.

B-line A particular genotype can be transferred to a male sterile line by continuously using the line with that genotype as the pollinator or 'B Line' male sterile lines developed by this method will contain the genes of the backcross parent in the cytoplasm of a non-recurrent parent.

Blood flower Fam. Asclepidiaceae, *Asclepias* sp.

Blumea lacera Fam. Compositeae.

Body In *Funaria* (Bryophyta), the body or theca is the spore containing part of the capsule and shows much internal differentiation.

Body cell The pollen grain in *Pinus* (Gymnosperms) at germination, the exine ruptures and entine grows out to form pollen tube which makes its way into the tissue of the megasporangium. The tube nucleus advances with the tip of the pollen tube. The generative cell divides into a stack cell and a body cell.

***Boehmeria* sp.** Rhea or Ramie, an important filera plant.

Boerhavvia diffusa Fam. Nycteginaceae. A common weed. Medicinal plant.

Bog An area of wet spongy ground consisting of decomposing moss and other vegetation. It often form with the growth of moss especially Sphagnum on the surface of a shallow pond or lake. Decaying moss and other vegetable matter will gradually accumulate in filling the pond to produce a quacking bog. With further compaction, peat will ultimately form.

***Bombax* sp.** Fam. Bombacaceae. Red of silk cotton tree.

Borassus flabekkifer Fam. Amaryllidaceae. Palmyra palm.

Border parenchyma In the entire vascular bundle of *Angiosperms i.e.* xylem, phloem and sclerenchyma is surrounded by a layer of large and thin walled parenchyma cell–the bundle sheath or border parenchyma.

Botany Biology is divided in two branches, namely Botany–that deals with the study of plant life, and Zoology–that deals with the animal life.

Bottom recessive An individual homozygous for the recessive alleles of all the genes under investigation.

Bougainvilles These are ornamental plants which bear floral bracts of many colours. The bracts are brightly coloured and are sometimes mistaken for petals.

Bracts These are the special leaves from the axil of which a solitary flower, or a cluster of flowers arises.

Bracteole When a small leafy thin structure is present on any part of the flower stalk (pedicel), it is called bracteole.

Bract scale In Gymnosperms, in female stroblii or cone bears a number of paired scales which are spirally arranged and tightly packed. In each pair, the lower scale is small and membranous, is called scale or carpellary scale.

Branching The mode of arrangement of the branches on the stem is known as branching. These are two principal types of branching *viz.* lateral and dichotomous.

Brassica compestris Fam. Crucifereae. An important oilseed plant.

Breeder seed Seed produced by the breeder and used to produce foundation seed.

Breeding The science of changing plants and animals genetically.

Brinjal Fam. Solanaceae. *Solanum melongena.* An important vegetable plant.

Bristles These are short stiff, needle like hairs, usually growing in clusters and not infrequently barbed.

Broomrape A root parasite of crop plants. *Orobanche Carnua* and *A. indica.*

Bryophyllum pinnatum The sprout leaf plant. A series of adventitious buds (foliar) are produced on the leaf margin, each at the end of the vein.

Bryophyta (1) These are very simple plants but they are more developed than the thallophyta. They do not possess any root but have characteristically thread like structures–the rhizoids which fix the plants with the substratum and help in absorption. There are no vascular tissue in the group. (2) A division of Crytogams, in which the plant body is thalloid or leafy; there is regular alternation of generation; the main body is always a gametophyte; the sporophyte always grows attached to the gametophyte as a dependent body.

Buckwheat *Fagopyrum* sp.

Bud It is an undeveloped shoot consisting of a short stem and a number of tender leaves arching over the growing apex. The method of outgrowth formation for production of new individuals in yeast is also called budding.

Budding (1) A mode asexual reproduction in which new organisms develop from the parent body in the form of an outgrowth. (2) This is a method of formation of new cells by division. This is characteristic of the yeast fungus which is unicellular. Also *See* Crop Science.

In Saccharomyces yeasts, they are also called budding yeasts.

Bud sport A somatic situation or variation arising in a bud and producing an abnormal branch.

Bud mutation If a mutation occurs in the meristematic tissue, in a very early stage of bud development, all the cells of the bud will be mutant in nature and the bud develops into a shoot, all the cells of the shoot will be of the mutant type.

Bulb An underground modified shoot (a really single, often large, terminal bud) consisting of a shortened convex or slightly conical stem, a terminal bud and numerous scale leaves (which are the swollen bases of foliage leaves) with a cluster of fibrous roots at the base.

Bulbils (1) When some of the smaller flowers of the inflorescence become modified into small multicellular reproductive bodies, these are called bulbis. (2) These are small resting buds developed on the rhizoids or protonema in Bryophytes. With the onset of favourable conditions, these buds or bulbils grow to form protonemal filaments which bear a crop of leafy shoots.

Cycas (Gymnosperm) reproduces vegetatively by special vegetative buds called bulbils.

Bulk breeding Growing of genetically diverse population of self pollinated crops in a bulk plot with or without mass selection followed by a single plant collection.

Bulk method of selection The F_2 plants are harvested in bulk and a random sample of the seed is taken for the sowing of the next generation. This procedure is repeated till the F_9. For single plant, selections are made in the F_{10} generation and progeny rows are raised in the $F_{11.}$ Desired rows are selected in the F_{11} and advanced to yield trials in the next year.

Bundle sheath Same as border parenchyma.

Butcher's broom Ruscus. An ornamental plant. Fam. Liciaceae.

Butter cup Fam. Ranunculaeae. *Ranunculas sp.*

Butterfly pea Fam. Papilionaceae. *Clitoria ternatea.*

Cabbage Fam. Crucifereae. *Brassica oleracea* var. capitata. An important leafy vegetable plant.

Cactus This is a xerophytic plant suited for the arid climate and sandy soils. Such plants have to guard themselves against excessive evaporation of water; which they do by reducing the evaporating surfaces. They have also to adopt special mechanisms for absorbing moisture from the soil and sub soil and retain it.

Caducous Such stipules which shed before the leaf unfolds.

Caemotaxis Attraction or repulsion of organisms by a diffusing substance.

Caesalpinia bonducella Fam. Caesalpinieae. Fever nut, of medicinal value.

Caesalpineae A sub family of Leguminoseae.

Caesalpinioideae The Cassia family. The family includes about 133 genera and is distributed in the warmer parts of the world. The floral formula is :

$$\oplus, ♂, K_5, C_5, A_{3+3+3}\ \underline{G}_1$$

Cajanus cajan Fam. Papilionaceae. An important pulse crop.

***Caladium* sp.** A multicoloured aroid plant which is a good example of mimicry in plants. It gives an appearance of a snake and thus is saved from other animals.

Calculations of linkage by additive method The formula applied is

$$P_2 = \frac{E - M}{N}$$

where P = linkage value, E = sum of ends classes, M = sum of middle classes; and N = total number of progeny.

Callose The tissue functions actively during the growing period and almost cease to function during winter. At the close of the growing season, the sieve plant is covered by a deposit of a colourless shining substance in the form of a pad callos or callus pad.

Callus In spring when the active season begins, the callus gets dissolved. In old sieve tubes, the callus forms a permanent deposit.

Calvin cycle Calvin and his associates in 1957 discovered the complete sequence of reactions taking place in CO_2 function to synthesise carbohydrate. They showed it to be a cycle pattern of reactions and it now known as Calvin Cycle. The first compound was called glycenide of iron etc.

Calyptra (1) In Marchantia (Bryophyte), along with the growing of the other parts of archegonium, the wall of the venter grows and forms the calyptra which surrounds the capsule, the neck withers and disappears. (2) In *Riccia* (Bryophyta), the oospore grows, divides and re-divides to give rise to a small globular structure called sporogonium. Along with the development of the sporogonium, the ventre of the archegonium also becomes meristematic and forms a two celled thick protective covering known as Calyptra.

Calyptrogen The single layered root dermatogen, the apex merges into periblem; outside this the dermatogen cuts of many new cells forming a small celled tissues called calyptrogen.

Calyx (1) It is the lowermost whorl of the flower and consists of a number of green leafy sepals. (2) In Angiosperms, the outermost whorl is of small green (rarely coloured) leaf like structure called sepals collectively known as Calyx.

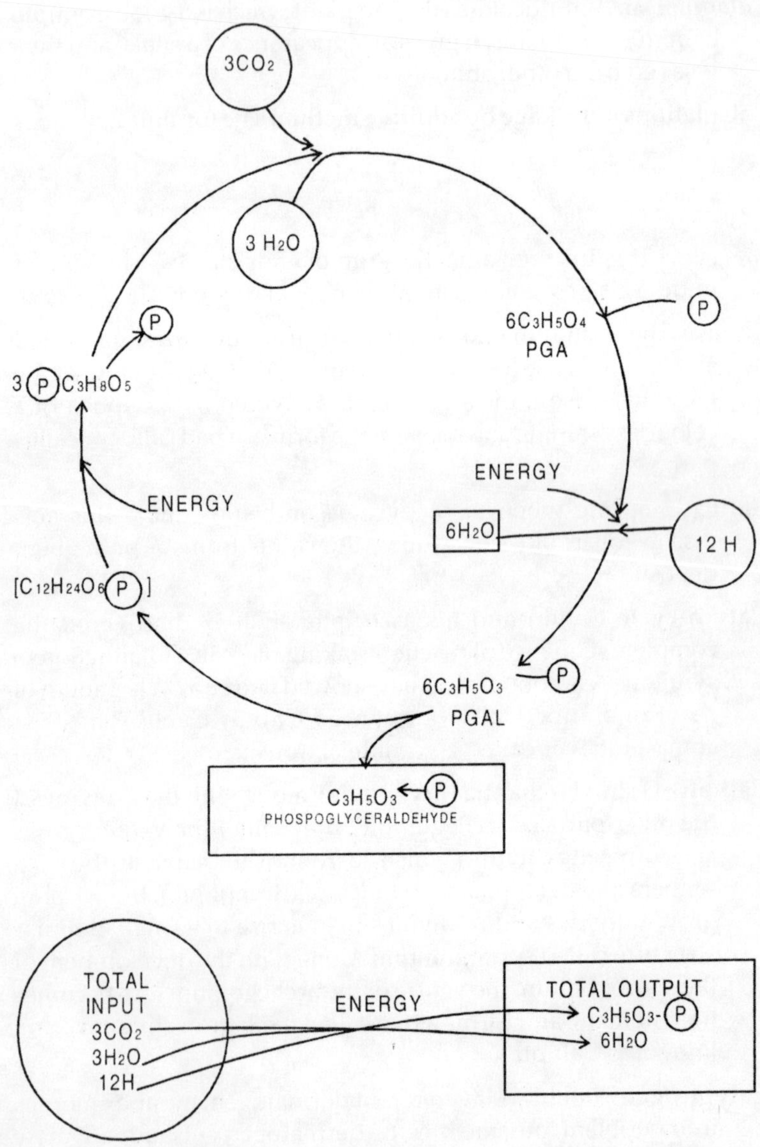

The Calvin Cycle

Summary of CO_2 fixation as a whole. The reaction cycle is shown above and the total input and output are below.

Cambium (1) in a cross section of dicot stem, passing inwards a band of thin walled tissue is seen, lying in between phloem and xylem; this is the cambium. (2) In the stems of flowering plants, the histological differentiation starts from both sides (inner and outer) of the procambial strand and in dicors a strip of meristematic cells is left in the middle known as Cambium.

Camellia The old genus for Thea to which tea belongs. The two species–*T. assimica* and *T. sinensis* are the common species grown for production of tea.

Camel's foot climber Fam. Caeselpinieae *Bauhinia vahlii.* Its long pods, sometimes as long as 30 cms., explode with a loud noise like that of a cracker, scattering the seeds in all directions. The species *B. purpurea* and *B. variegata* are ornamental trees.

Campylotropous When the transverse ovule is bent round like a horse shoe so that the micropyle and the chalaza do not lie on the same straight line. This position is also called curved position of the ovule.

Candytuft Fam. Crucifereae. *Iberis* sp. An important ornamental shrub.

Cane Fam. Palmacae. *Calamus* sp.

Cane sugar Sugar is extracted from sugar-cane. Fam. Graminae, *Saccharum officinarum.*

Cannabis sativas Fam. Leguminoseae. Hemp. Important green manure crop and give useful fibre also.

Capitulum An inflorescence in which the main axis is flattened, the receptacle supressed, becoming almost flat, and it bears a mass of small sessile flowers (florets) on its surface, with one or more whorls of bracts on the base forming an involucre.

***Capsella* sp.** Fam. Cruciferae. The shepherds' purse.

Capsicum annuum Fam. Solanaceae. The red pepper or chilli. Used as a condiment.

Capsule This is a dry, one to many chambered fruit developing from a syncarpous pistil, and dehiscing in several ways.

The bacterial cells are surrounded by a slimy gelatinous layer which is protective in function. In some bacteria, this layer forms a distinct envelope called capsule.

In *Funaria* (Bryophyta), the capsule is borne a little obliquely at the extreme of seta, is almost pear shaped in outline and is distinguishable into three well defined regions: (1) Apophysis, (2) Body or theca, and (3) Operculum.

Carambola Fam. Apocynaceae, *Averroha carambola.* An important fruit plant. Fruits rich in vitamin C.

Caraway Fam. Umbelliferae. *Carum carvi.* An important medicinal and condiment plant.

Carbohydrates Macromolecular organic compounds having hydrogen and oxygen in 2:1 ratio *e.g.* sugars, starch, cellulose etc.

Carbonaceous Denoting sedimentary deposits formed principally from remains of plants and other organic materials *e.g.* coal, peat and petroleum.

Carbonation A process of chemical weathering of rocks. It is produced by rainwater containing small amounts of carbon dioxide, from the atmosphere or soil on solution (weak carbonic acid) and is specially effective on limestones, composed largely of calcium carbonate, which is converted to calcium bicarbonate and removed in solution. $CaCO_3 + H_2O = Ca(HCO_3)_2$. The process is the chief agent of erosion in a limestone country.

Carbon cycle The circulation of carbon through the atmosphere, the oceans and the earth's surface. Carbon is stored in living organisms such as vegetation and also in the form of carbonate minerals (*e.g.* limestones) and fossil fuels. It is released in the atmosphere as carbon dioxide chiefly through the action of living organisms on land and in the oceans. Smaller amounts are also released through the weathering of the carbonate minerals, the decay of organic elements in the soil, the burning of fossil fuels (*e.g.* coal) and volcanic activity. Carbon dioxide can be returned to the biosphere through photosynthesis by plants, in the oceans carbonate of lime is ultimately formed from the shells and skeletons of the marine creatures.

Cardamom Fam. Liliaceae. *Elletteria cardomum.* An important spice plant.

Cardenthera triflora An amphibious plant showing heterophylly.

***Cardospermum* sp.** Balloon vine.

Carica papaya Fam. Caricacae. Papaya. An important fruit plant.

Carina The two innermost petals of the papilinacious flower, also known as keel.

Carnivorous plants (1) These plants are known to capture the lower animals of various kinds, particularly insects. They digest the prey and absorb proteins from their body. (2) These are those plants which possess special organs–the traps to capture and digest small animals. These plants grow in wet places like swamps and water logged areas. In these places, there are no bacterial activities to decompose the organic matter and these areas are poorer in nitrogen content.

Carotene It is a plant pigment, is the source of vitamin A and animals can synthesise it in their bodies by taking food of carotene containing plants. Plants like carrots, tomato, mango, orange contain this pigment.

Carotenoid Any of a group of yellow, orange or red pigments associated with chlorophylls in plants and in animal fat.

Caropophore When the thalamus extends up towards into a slender axis with the carpels remaining attached to it at first and separating from it afterwards on maturity. Such an axis is called carpophore.

Carpel (1) The female part of a flower. It is a part of the gynoecium of a flower whorl. (2) Carpel is the basic unit of gynaecium and is flask shaped in outline. The basal swollen portion is called ovary, long or short neck like portion is called style and the tip of the style is disc shaped stigma.

Carpellary scale In Gynmosperms. *See* bract scale.

Carrier molecule Pfeffer (1900) while planning the movement of water in the plant cells thought that chemical combinations with cell constituents may be involved in these processes and thus he conceived the idea of carrier molecules.

Carrisa carandus Fam. Apocynacae. An important fruit plant, fruits rich in vitamin C.

Carrot Fam. Umbelliferae. *Daucus carota.* An important vegetable crop plant, rich in vitamin A.

Carthamus tinctorius Fam. Compositeae safflower. An important oilseed.

Carum carvi Fam. Umbelliferae. Caraway. An important condiment crop plant.

Caruncle At one end of the seed coat, there is a white body and outgrowth formed at the micropyle. In some seeds as castor, small fleshy outgrowth is developed at the micropyle end. Such a small outgrowth is called caruncle.

Caryoata urens Fam. Palmae. Sago palm or Toddy palm.

Caryophyllaceous This form of corolla consists of five petals with comparatively long claws and the limbs of the petals are placed at right angles to the claws.

Caryopsis This is very small dry, one seeded fruit developing from a single pistil, with the pericarp fused with the seed coat.

Cashew nut *Anacardium* sp. Fam. Anacardiaceae. It is an important dry edible fruit.

Casparian strips Sometimes thickening in the endodermal cells is in the form of distinct strips or bands called casparian strips.

Cassia fistula Fam. Caesalpinieae. Ornamental tree.

***Cassystha* sp.** A stem parasite plant.

Castor Fam. Euphorbiaceae. *Ricinus communis.* An important oilseed of which oil is of medicinal value.

***Casuarina* sp.** Because of a strong development of cuticle, it can resist transpiration to a great extent.

Catabolism The processes that lead to the destruction or the breaking down of the synthesised food material is called catabolism.

Catalyst Any substance that accelerates the rate of chemical reaction but is not affected by itself in the process.

Catastrophism The religious people advocate that after each catastrophy God created a new world. Cuvier, a palaeontologist and contemporary of Lamark concluded that earth has experienced 27 catastrophies.

Catechu Fam. Mimoseae, *Acacia catechu.* Yields a kind of tannin obtained by boiling the heart wood.

Catkin This is spike with a long and pendulous axis which bears unisexual flowers only.

Caudate When the apex is drawn out into a long slender tail.

Caudex The unbranched cylindrical and stout stem, marked with scars of fallen leaves is called caudex.

Cauliflower Fam. Cruciferae, *Brassica oleracea* var. botrytis. The curd head is used as a vegetable.

***Ceiba* sp.** Fam. Malvacae. The white cotton tree.

Cell The structure unit of living organisms, consisting basically of nucleus, cytoplasm and in plants a cell wall.

Cell membrane Next to cell wall is the cell membrane. Leaving aside the cell wall, the membrane forms the boundary of the cells. The cell is enclosed in a semipermeable membrane the cell membrane (or the plasma membrane), through which metabolic exchange occurs. A cell wall often made up of cellulose is found in addition in the plant cell.

Cell permeability The presence of mineral salts in the plant cells is a proof of the fact that these substances can pass through the plasma membrane. The ability of the protoplasm to allow the diffusion of dissolved substances through it is known as permeability. While the plasma membrane is permeable to some solutes, it is impermeable to others.

Cell plate A cell plate is formed within the spindle at the equatorial region. This increases in diameter until it reaches the lateral walls and separates the two nuclei from each other.

Cellulose The wall of young cell is composed of cellulose which is an insoluble carbohydrate.

Cell wall It is the non-living part of the cell and is secreted by the protoplasm. It gives the protoplast a definite shape and constitutes the structural frame work of the plant. Cellulose is the main conspicuous substance of which the cell wall is made up. The cells which are to differentiate into or trachieds show characteristic thickening of the cell wall.

Celsius scale A temperature scale in which temperate of melting ice is taken as 0°C and the temperature of boiling water is 100°C. In meteorology, the scale is more commonly known as the centigrade scale.

Censer mechanism *See* Aristolochia gigas.

Centella asiatica Fam. Umbelifereae. Indian pennywort.

Centimorgan An unit of genetic map distance equal to the distance along a chromosome that gives a recombination frequency of one parent.

Central body In Oscillatoria (algae), the protoplasm of each of the filament is differentiated into two regions–a coloured peripheral zone-the chromoplasm and an inner colourless zone-the central body.

Centre of origin From a study of crop plants, variations and their wild allies, Vavilov found that there are some regions which are the centres of genetic diversity. These regions are rich in wild relatives of crop plants and also in various cultivated crops.

Centric leaf When the leaf is more or less cylindrical and directed upwards or downwards.

Centromere A non-staining localised region in each chromosomes to which the spindle fibre appears to be attached at metaphase. It is the centre of movement of chromosome and at anaphase leads the way when the chromosomes or the chromotids pass to the poles.

Centroplasm In Algae, protoplasm is differentiated into two regions; an outer pigmented one known as chromoplasm and a central colourless centroplasm.

Centrosome (1) A self propagating cytoplasmic body present in animal cells and some lower plant cells, consisting of centrioles and astral rays at each pole of the spindle during nuclear division.

(2) In the neuromotor apparatus of algae, the two basal granules are connected with each other by a thread like structure paradesmos. The thread is further connected with the granular centrosome by another thread like process called rhizoplast.

***Ceratophyllum* sp.** Hornwort.

***Cercus* sp.** The oak tree.

Cereal A cultivated grass used as human and animal food. The most important crop in terms of human consumption is rice which is the staple diet of over one quarter of the world's population.

***Ceriops* sp.** A mangrove plant.

Certified seed Seed used for commercial crop production produced from foundation registered, or certified seed, by an official certifying agency.

Cestrum nocturnum Fam. Solanacae. An ornamental plant used for its sweet small.

Chalaza The basal portion of the swollen nucellus in an angiosperm ovule. The distal end of raphe or the funicle which is the junction of the integuments and the nucleus.

Chalazogamy During fertilization in angiosperms, the pollen tube enters the body of the ovule *i.e.* nucellus to reach embryo sac. The pollen tube may enter the ovule through micropyle when it is known as porogamy, or through chalazal end when it is known as chalazagamy.

Chalybeate Water impregnated iron salts. It is usually applied to mineral water or spring water.

Chantonastic When the plant organs characteristic movements irrespective of the direction of the stimuli but are influenced by diffused stimuli, they are known as nastic movements. They are of three kinds–(1) Nyctinnastic, (2) Chemonastic, and (3) Seismonastic. Floral parts and leaves take different direction by day and night. Such diurnal movements are known as nyctinastic movements or sleeping movements. The movements induced by a chemical substance are known as chemonastic. The movement induced by mechanical stimuli are known as seismonastic movements.

Character or characteristic An attribute of an organism resulting from an interaction of a gene or genes with the environment.

Chemical mutagens Oehlkers in 1943 discovered that a mixture of ethylurethane and potassium chloride induced translocation in Oenothera and by Auerbach and Robson in 1946 that mustard gas $(ClCH_2)_2S$ produced mutation in Drosophila, it was felt that mutations could be produced by chemicals also.

Chemical nature of protoplasm The composition of protoplasm is always changing because of several activities going on in it, and no two analyses give identical results. Protoplasm always contains some non-living substances which are the products of its metabolic activities. Whenever an attempt is made to analyse the protoplasm chemically, it is killed so that analyses is of the dead protoplasm and not of the living one is obtained.

Chemistry of respiration There are two major phases of respiration–(1) the oxidation of carbohydrates to pyruvic acid–a process called glycosis, and (2) the subsequent oxidation of pyruvic acid–the Kreb's cycle.

Chemosynthesis Certain bacteria get energy from oxidation of inorganic substances and utilise this energy for carbon assimilation. Reduction of carbon dioxide by this method where instead of radiant energy chemical energy is utilised as chemotherapy.

Chemosynthetic bacteria *See* autotrophic bacteria.

Chemotaxis It is the movement in response of chemical substances. The antheroizoids in ferns and mosses are attracted towards the archegonium by chemical substances secreted by them.

Chenopodium album Fam. Chenopodiaceae. Have medicinal value. It is weed of widespread occurrence.

Chestnut (water) It has a submerged stem. The important fruit of *Trapa* sp. Bears edible fruits.

Chlamydomonus An algae with about 400 species is cospmopolitan in distribution. It is the simplest unicellular green alga commonly found in stagnant water of ponds, pools, ditches, puddles etc.

Chlamydospores In fungi *(Mucor recemosa)* if the conditions become unfavourable, the oidia become thick walled and are known as chlamydospore.

Chlorenchyma When parenchyma cells contain chloroplasts, it is termed as chlorenchyma. It provides the site for photosynthesis.

Chloronemal branches In (Bryophyta) *Funaria*, the protonema is very much similar to the filamentous alga *Coonferva* from where the term confervind is derived. The protonema usually has two types of filaments–ordinary green filaments called chloronemal branches or chloronema and colourless rhizoidal filaments.

Chloroplasts The plastids are protoplasmic bodies of various shapes present in the cytoplasm. They are usually spherical, oval, ellipsoidal or discoid. Each plastid has a ground substance called stroma in which are embedded a large number of pigment granules. The plastids are derived from proplastids transmitted materially. The plastids are known as leucoplasts when they are colourless, as chloroplasts when they are green and as chromoplasts when they are of other colours. This is essential for photosynthesis; the plastids are powerless in this respect without the presence of chlorophyll. For the same reason, non-green parts of plants cannot photosynthesis.

Chiasma (pl. chiasmata) The point of contact and interchange of partners between paird chromatids observed between diplonema and II metaphase in meiosis.

Chicory Fam. Compositeae. *Chichorium intybus.*

Chilli Fam. Solanaceae. *Capsicum annuum.* Bears vegetable fruits.

Chimera If a mutation occurs in the later stages of bud development, only some of the cells of the bud will be mutant in nature and when the bud develops into a shoot, some of the cells of shoot will be of the mutant type and others will be of the nonmutant type. A plant which has genotypically distinct tissues lying adjacent to one another is called a chimaera or chimera.

China rose Fam. Malvaceae. *Hibiscus rosa-sinensis.* The shoe flower.

Chlamydomonas It is a unicellular green algae found in ponds, ditches and other pools of stagnant water.

Chloritic In the absence of iron, magnesium and manganese in the soil, the plants tend to be chlorotic *i.e.* they are not able to form chlorophyll, the green matter of plants.

Chlorophyll It is not one simple substance but a mixture of four different pigments, *viz.* chlorophyll-a (blue black), chlorophyll-b (green black), carotene (orange red) and xanthophyll (yellow). Chlorophyll-a and chlorophyll-b are associated with each other in the chloroplast, but carotene and xanthophyll may also occur without chiroplast in any part of plant also. Chlorophyll-a is $C_{55}H_{72}O_5N_4MG$; Chlorophyll-b is $C_{55}H_{70}O_6N_4MG$; Carotene is $C_{49}H_{36}$ and Xanthophyll is $C_{40}H_{36}O_2$.

Chlorophyllous tissue This tissue consists of one or two layers of chloroplasts bearing cells, intruding inwards here and there.

Chloroplasts Plant cells contain plastids of various kinds, the important of which are the chloroplasts. These are centre of photosynthesis activity.

***Chorchorus* spp.** Fam. Tiliaceae. The best fibre of this plant (*C. capsularis* and *C. olitorius*) is widely cultivated in the low lying areas of eastern India. It is economically important fibre plant.

Chromatic reticulum The. main mass of the nucleus in a cell not undergoing division appears as a fine network of threads constituting the chromatin reticulum.

Chromatid One of the two longitudinal halves of a chromosome which shares a common centromers with a sister chromatid; results from the replication of chromosomes during a nuclear division.

Chromatin Nuclear material (DNA, RNA, histone and non-histone proteins) composing the chromosomes. It is a densely staining substance in the nucleus.

Chromocentre Irregular mass of heterochromatin.

Chromomere The chromatin granules of characteristics size and position, small stainable thickenings arranged linearly along a chromosome.

Chromonema (pl. chromonemata) One of the delicate helically coiled thread-like filament composing a chromosome. It is a optical single thread within a chromosome.

Chromoplasts These are variously coloured plastids–yellow, orange and red. They are mostly present in the petals of flowers, fruits and the colouring matter associated with them.

Chromosomal aberrations Chromosomes are structures of definite organisation but when breaks occur in chromosomes, changes in their structure follow. Radiations and chemicals can also cause breaks in chromosomes causing lot of genetic changes.

Chromosomal deficiency A deficiency involves the detachment and loss of a portion of an arm from the remainders of the chromosomes.

Chromosome Nucleoprotein structures, generally more or less rod-like during nuclear division, the physical sites of nuclear genes which are arranged in linear order. Each species has a characteristic number of chromosomes, although individuals with fewer or more than the characteristic number occur, especially in plants.

Chromosome complement A collective term for the group of chromosomes in single nucleus whether somatic or gametic.

Chromosome hybrid sterility Sterility caused by inability of homologous chromosomes to pair during meiosis due to chromosomal aberrations.

Chromosome map A diagram of the relative position of the genes in a chromosome. Same as linkage map.

Chromosome sterility Sterility in autopolyploids, interspecific hybrids, aneuploids and individuals carrying chromosomal aberrations is very often due to chromosomes.

Chromosome set The whole complement of chromosomes present in, or derived from, the nucleus of a gamete.

Chromosome theory The regular and precise longitudinal division of the chromosomes and the reduction in the number of chromosomes from the diploid to the haploid state (n) during the information of gametes by meiosis and restoration of the diploid number of chromosomes in the zygote by fertilisation showed that the chromosomes are of great importance to the cell. Even before Mendel's work was rediscovered, the existence of Mendelian units known, it was realised that chromosomes are intimately concerned with inheritance.

Chrysopogan acicilata Fam. Gramineae. Love thorn. Important grass of pastures.

Cicer arietenum Fam. Papilionaceae. Gram (horse). An important pulse crop of tropics.

***Cinchona* sp.** Fam. Rubiceae. Yields quinine and ipecac medicines.

***Cinnamomum* sp.** An oil bearing plant, medicinal value.

Circinate The leaves of ferns are pinnately compound and circinate *i.e.* rolled from the apex downards. The young leaves of *Cycus* show circinate ptyxis as in ferns.

Circulation When the protoplasm moves or streams in different directions within a cell in the form of delicate strands round a number of small vacuoles, the movement is called circulation.

Circumnutation It is the movement of growing organs due to unequal growth on different sides of the organ and is termed as nutation. If the growth takes place in a regular sequence around the stem-apex in a spiral form as in the case of tendrils, the movement is known as circumnutation. In leaves, the growth is at first more rapid on the lower side (hyponasty) and later on the upper side (epinasty).

Cissus quadrangularis The wind vine. Fam. Vitaceae.

Cistron A segment of DNA specifying one polypeptide chain in protein synthesis. Under the concept of triplet code, on cistron must contain three times as many nucleotides pairs as many amino acids in the chain it specifies.

Citric acid cycle During aerobic phase of cell respiration, with the access of O_2 the pyruvic acid is completely oxidised by a group of enzymes called oxidases into CO_2 and H_2O. Maximum amount of energy is released by this process. A series of reactions takes place in a cyclic order, called citric acid cycle or Krebs cycle.

Citrullus lanatus Family Cucurbitacae. Water melon. An important edible fruit.

Citrullus vulgaris Though the water melon is believed to be a native of tropical Africa, it is cultivated extensively in Japan, Philippines, China, India and USA. The triploid seedless watermelons are very popular.

Citrus fruits Varieties of topical and subtropical fruits that contain citric acid of which lemons, lime and oranges are most common.

***Citrus* spp.** Fam. Rutacae. Includes all citrus fruits like lime, lemon, citron, pummelo, grape fruit, oranges, mandarins.

Cladode In some plants, one more short, green cylindrical or sometimes flattened branches of limited growth develop from the node of the stem or branch in the axil of a scale leaf. Such a branch is the cladode.

***Clamatis* sp.** Fam. Ranunculaceae. A small tuberous plant with wooly achenes for wind dispersal. Virgina bower.

Clausena pentaphylla Fam. Rutaceae.

Claw In caryophyllaceous form of corolla, the five petals with comparatively long claws and limbs, are arranged in the form of cross.

Cleistogamy In such cases, the bisexual flowers never open. The flowers remaining closed, the pollen grains only pollinate the stigma of the same flower.

Cleistothecium It is the body, a closed fruiting body after the fusion of two nuclei in the ascus (fungi). After the plasmogamy and the formation of asci, the stalk cells of both the sex organs branch freely to produce a large number of sterile hyphae which coil themselves around the archicarp to form a distinct sheath called peridium. The whole structure enclosed in the peridium is known as fruit body or ascocarp. The ascocarp along with peridium forms cleistothecium.

Climax vegetation The form of vegetation that has become established in response to natural conditions prevailing in a particular environment. It is a final achieved over many years of the natural processes of succession within the plant community. Climax vegetation is governed most strongly by regional climatic factors and hence is often called Climatic climax vegetation.

Climbers These plants develop special structures of attachment which they cling to the neighbouring objects for the support of their body and for aid in climbing. There are rootlet climbers, hook climbers, tendril climbers, leaf climbers, etc.

Clinostat This equipment is used to demonstrate geotropism.

Clitoria terneta Fam. Papilionaceae. Butterfly pea.

Clonal culture A culture in which all the organisms present have arisen from a single individual of isogamous species, sexual reproduction never occurred unless they were mixed with others of opposite mating type.

Clonal selection Species that do not at all produce seeds *e.g.* ginger, turmeric, etc. or that produce seeds very rarely *e.g.* banana or that produce seeds only under special conditions *e.g.* sugarcane, are propagated asexually. Vegetative propagation is also resorted to in fruit trees like mangoes, apples, oranges etc. to maintain the uniformity of variety. Thus the sugarcane is propagated by sets, grapes by cuttings, grasses by slips, potatoes by tubers, onions by bulbs, banana by suckers, mangoes by grafts and roses by buds. A variety propagated asexually from a single individual is known as clone.

Clone A group of plants descended by asexual reproduction (mitosis) from a common ancestor.

***Clostridium* sp.** These are anaerobic bacteria in soil, having the power of fixing the free nitrogen of the soil air in their own bodies in the form of amino acids and finally building proteins from them.

Cloves The dried flower buds of *Syzygium aromaticum*, the green colour of bud turns to dark brown on drying. Cloves are extensively used in curries, preserves and medicines.

C-mitosis The type of abnormal mitosis, produced by Colchicine treatment, in which the spindle is inactivated and the chromosomes become scattered in the cytoplasm.

Cobalt^{60} It is an important source of pure gamma-rays of high energy. This is radioactive isotope of Cobalt (Co).

Cocci These are spherical bacteria *e.g.* Styphylococcus, Streptococcus, Diplococcus, Micrococcus, Azotobacter, etc.

Coccinia indica Fam. Cucurbitaceae. It is an important vegetable.

Coccus In these type of bacteria, cells are spherical or globular. The cocci have further been divided into (1) diplococcus–cells often joined in pairs, (2) tetracoccus–tetrad of four cells, (3) steptrococcus–filaments of cocci. They are formed by the repeated divisions of the parent cell in the same plane.

Cocoa It is prepared from the seeds of *Theobroma cocoa*, a small tree. Each tree commonly bears 70–80 fruits.

***Cocoloba* sp.** A cactus plant.

Coconut Fam. Palmae *Coconut nucifera*, an important oilseed having medicinal and industrial value.

Codon A set of nucleotides that is specific for a particular amino acid in protein synthesis; generally agreed to consist of three nucleotides and in the case of some amino acids, may be any of the four nucleotides.

Coenobium In *Volvox* (alga), the multicellular motile colony with definite number of individuals is called coenobium.

Coenocyte Mass of cytoplasm containing many nuclei, formed as a result of repeated division of the nucleus without the division of cytoplasm.

Coenotytic There is usually a single nucleus in a cell, but sometimes there may be more than one nucleus in a single cell *e.g.* fungi and endosperm of ovules in early stages of development. A cell with more than one nucleus in it is termed as coenocyte or coenocytic.

Coenzyme An organic compound which activates an enzyme.

Coffee This is a favourite for many. It is prepared from the seeds of *Coffee arabica* and *C. robusta*, particularly the former, are the source of coffee.

Coinvalia ensiformis Fam. Leguminoseae. Sword bean. Important vegetable plant.

Coir It is the husk fibre of dry coconut fruits, the fibres are largely used for making mats, mattings, mattresses etc.

Colchicine This chemical brings about doubling of chromosomes by acting on the spindle mechanism. The two chromatids of each chromosome separate at the centromere but do not go to the poles. A nuclear membrane is formed around the whole group of chromosomes resulting in the formation of a single nucleus with 4×2 chromosomes, C-mitosis.

Colchicum autumnale The alkaloid colchicine is derived from the corms and seeds of the autumn crocus, of family Liliaceae, it is very effective in doubling the chromosome number, an easy and effective method for production of polyploids is available to plant breeders.

Coleoptile At seed germination, while the plumule breaks through the upper distinct, cylindrical portion of the plumule sheath. The sheath is the coleoptile.

Coleorhiza In the germination of the monocotyledonous seeds, like paddy and maize, the radical makes its way through the lower short collars like end of the sheath called root sheath or coleorhiza.

***Colias* sp.** It is called yellow clover butterfly. In this, species, females may be either yellow or white but males are always yellow, white colour in the females depneds on a dominant gene W so that females of genotype WW or Ww are white while ww are yellow. Males are yellow irrespective of whether they are WW, Ww or ww. The genotypes of the females can be determined by mating them with yellow females and observing the colour of the female progeny. If all the females progeny are white, proportion of 1 white: 1 yellow, the male parent is Ww. If all the female progeny are yellow, the male parent is ww. Thus white colour is a sex-limited character.

***Colleus* sp.** Fam. Labiatae. It is a plant used as an ornamental shrub.

Collenchyma (1) This tissue consists of somewhat elongated cells with the corners or the intercellular spaces much thickened with a deposit of cellulose and pectin. (2) Such cells are somewhat elongated with oblique end walls. In transverse section, they appear to be rounded oval or polygonal. The cells are much thickened at the corners against the intercellular space, the middle portion of the wall remain unthickened.

Collinearity The state in which the sequence of nucleotide in a cistron corresponds with the order of the amino acids in the polypeptide it specifies.

Colloid A chemical substance having particles which range from 1 m to 100 m in size.

***Colocasia* sp.** The underground stem is used as vegetable.

Colocynth Fam. Cucurbitaceae. *Citrullus colocynth.* The fruits have medicinal value.

Columella In the capsule of Bryophytes, it is the solid central column. In the development of sporangia in the fungi *(Mucor),* the central vacuolated region is separated from the outer region by a number of small flattened vacuoles, which gradually become more flattened and merge with one another to form a distinct cleft which separates the central dome-shaped sterile region called columella.

Combining ability (general) The average performance of a strain in a series of crosses.

Combining ability (specific) The performance of specific combinations of strains in crosses.

Combining inbreds into double crosses The double cross is a cross of two single cross hybrids (A × B) × (C × D). Each single cross hybrid is the F_1 of a cross between two inbred lines and is heterozygous but productive, and gives seeds in abundance. It is these seeds that are distributed to the farmers for growing plants in the next season.

Combining inbreds into single crosses A sing cross is a cross between two inbreds (A × B). Since the inbreds used in a single cross are homozygous, the single cross progeny are heterozygous, the single cross progeny are heterozygous all loci in which the two inbreds have different alleles. A superior single cross progeny is more vigorous and productive than the open-pollinated varieties from which the inbred lines are developed.

Commas These are bacteria which look like a comma (,).

Commelina bengalensis Fam. Commelineceae. Ciculation of protoplasm is very distinctly seen in the staminal hairs of this plant.

Companion cells Associated with each sieve tube and connected with is by simple pits, there is a thin walled, elongated cell, called the crompan cell. These are typically found in angiosperms but are absent in pteriophytes and gymnosperms. They are elongated narrow living cells and retain their nuclei even at maturity. Their protoplasm is continuous with the protoplasm of sieve tubes through the perforations and it is thought that their nuclei control the activity of the sieve tubes as well. They assist the sieve tubes in conduction and may store food material themselves.

Companulate When the shape of the corolla resembles that of a bell, it is said to be bell shaped or companulate.

Compensation point It is a point gas exchange and oxygen emission where respiration is equal to photosynthesis in terms of gaseous exchange. The light intensity where this point is reached is termed as compensation point.

Competence Ability of bacterial cells to incorporate DNA and become genetically transformed.

Complimentary cells A lenticel appears as a small oval or elliptical scar on the surface of the stem. A transverse section through the lenticel shows that at these places phellogen instead of producing normal cork cells towards outside produces oval to rounded thin walled parenchyma cells with intercellular spaces. These cells are known as the complimentary cells. Section through one of the scars shows that the lenticel consists of a loose mass of small thin walled cells, the complementary cells.

Complementary genes Non-allelic genes that act together to produce a phenotype different from that produced by either alone. It gives a ratio of 9:7 in the F_2 progeny.

Complete linkage Same as absolute linkae.

Compositeae Family of Angiosperms. The largest family of angiosperms containing about 950 genera and about 20,000 species. Its floral formula is:

$$\oplus, \male, K_2 \text{ (Scales) } C_5 A_5 \overline{G}$$

Compound A chemical substance formed by the chemical combination of two or more simple substances. Plant leaves are compound (Botany).

Concentric bundle When xylem lies in the centre and is surrounded by phloem or phloem lies in the centre and is surrounded by xylem. A concentric bundle is always closed.

Concordance Identity of matched pairs or groups for a given trait, both expressing the same genetic syndrome.

Conduction The water with the mineral salts absorbed from the soil by the root hairs gradually accumulates in the cortex. As a result, the cortical cells become fully turgid. Under this condition, their elastic walls being much stretched, exert pressure on the fluid contents and force out a quantity of them towards the xylem vessels, and the cortical cells become flaccid. They again absorb water and become turgid and the process continues.

Cone In cycads (Gymnosperms), are dioecious, *i.e.* male and female flowers are borne by two separate plants. The male flower is a cone borne at the apex of the stem. Side by side temporary association of two bodies; as of synapsed chromosomes in meiosis, or of two organisms during reproduction.

Conical root When the root is broad at the base and it gradually tapers towards the apex like a cone, it is said to be conical.

Conidia In fungi *(Albugo)*, the sporaniophores produce the sporangiospores called conidia. *See* Plant Pathology.

Coniferous forest A vegetation type dominated by trees whose seeds are carried in cones and which are chiefly evergreen. Coniferous forests usually consist of stands of a relatively few trees species, such as pine and spruce of uniform appearance. Their close growing nature and thẹ particular climate they grow in restricts undergrowth and the development of stratification.

Conjugating tube In *Ulothrix*, the protuberances grow further and push the pairing filaments a little apart. Finally the mutual walls between the protuberances get dissolved by enzymatic action so as to form a continuous passage called conjugating tube.

Conjugation In lower plants like algae and fungi, the pairing gametes are essentially similar *i.e.* not differentiated into male and female and are called isogametes. The union of such similar gametes is called the conjugation and the zygote thus formed is called zygospore. In bacteria, the sexual union of two bacteria of different strains as in *Echeriachia coli*, where two bacterial cells come to the side by side, the male bacterium injects its chromosomes into the female and subsequently dies off.

Conjunctive tissue The parenchyma lying between xylem and phloem bundles constitute the conjunctive tissues. Phloem bundles contain sieve tubes, companion cells and phloem parenchyma. Pith is normally absent and when present it is very small composed of only few parenchymatous cells lying in between and surrounding the bundles comprise the conjunctive tissue.

Connective Each stamen consists of filament, anther and connective. The anther lobes are joined together by a tissue called connective.

Consanginity Descent from a common ancestor.

Consanguineous mating Mating between relative in humans.

Continuous variation Those in which the variations differ from each other by infinitely small steps.

Contractile vacuole In *Chlamadomonas* (alga), there are two contractile vacuoles at the interior end at right angle to the base of the flagellum.

Contributing genes Same as active genes.

Control For testing the progeny lines in attesting procedure, a control (the local variety) is grown among the rows of the variety or varieties. A known variety used for comparing the performance of newly developed strains during the evaluation.

Convolvulaceae A family of angiosperms.

Convolvulus arvensis A common weed of cultivated crops.

***Convolvulus* spp.** The common weeds like railway creeper (C. spp.).

Copra The fleshy kernel of a coconut that has been dried. This is used in confectionery as desiccated coconut and is also crushed to produce coconut oil which is used in the preparation of cosmetics and soap.

Coral tree Family Papilionaceae. *Erythrina variegata*. In an ornamental tree.

Corculum Also known by the genus as Antigonon.

Cordate When the blade is heart-shaped as in betel *(Peperomia* sp.*)*.

***Cordyline* sp.** Fam. Liliaceae.

Coriandrum sativum Fam. Umbelliferae. It is an important spice coriander.

Cork The new cells cut off by the cork-cambium on its outer side are roughly rectangular in shape and soon become suberized. They form the cork of the plant.

Cork cambium Sooner or later another tissue *i.e.* the cork-cambium (or phallogen) makes its appearances in the cortical region. Commonly it originates in the outer layer of collenchyma.

Cork tree Margosa.

Corky Thickened and hardened group of dead cells.

Corm This is condensed form of rhizome and consists of a stout, solid, fleshy underground stem growing in the vertical direction. It is more or less round in shape.

Corn Maize. Fam. Gramineae. An important food crop.

Corolla It is the second whorl of the flower and consists of a number of usually brightly coloured petals. In a flower, inside the calyx is the second whorl called corolla (petals).

Corolloid roots In *Cycus* (Gymnosperms), the primary root is large especially in the seeding stage. It bears many lateral branches which in turn bear bunches of small dichomatously branched negatively geotrophic roots which look like corals and hence the name corolloid roots.

Corona Sometimes by the transverse splitting of the corolla, an additional whorl may be formed at its throat. This additional whorl may be made up of lobes, scales or hairs, free or united, is known as corona.

Cortex This is the zone that lies between the epidermis and the pericycle varying in thickness from a few of many layers. It lies inside the: epidermis and varies in different species. Internally it is limited by endodermis.

Corypha umbraculifera Fam. Palmae. Talipot palm, bears a huge inflorescence once only after 150 years and then dies.

Cosmos Fam. Compositeae. A ornamental herb.

***Cosmos* sp.** A kind of inhibition regularly occurs at the stigmatic surface, with the result the pollen do not at all germinate. This type of incompatibility is also seen in *Brassica oleracea* and *Raphanus sativus*.

Cotton Fam. Malvaceae. *Gossypium* spp. An important fibre plant.

Cotyledons The whole whitish body as seen after removing the seed coat, is the embryo. It consists of (*a*) two fleshy cotyledons or seed leaves and (*b*) a short curved axis to which the cotyledons are attached.

Coupling The stage of having two linked recessive genes on one chromosome and their two dominant alleles on the other.

Coupling (inheritance) It is a condition in linked inheritance in which an individual heterozygous for two pairs of genes received the two dominant members from one parent and two recessives from other parent.

Cowage *Mucuna* sp. It is a larger twiner. It bears stinging hairs of fruits.

Crawfoot (water) Fam. Ranunculaceae. Ranunculus aquatilis.

Cremocarp A type of fruit. It is dry, two chambered inferior fruit splitting into two indehiscent, one seeded pieces called mericarp.

Crenate When the teeth of the margin of leaf are rounded.

Crepe tree *Lagerstroemia* sp. The seeds bear thin membranous wings for facility of dispersal. An ornamental shrub.

Cress Garden cress. Fam. Cruciferae. *Lepidium sativum.*

Crick and Watson They shared the Nobel Prize in 1962 for elucidating the structure of DNA.

Crinum asiatica Fam. Amyryllidaceae.

Criss cross inheritance in Drosophilla studies, when a white eyed female was crossed with a red eyed male, it produced red eyed female and white eyed males. This method of inheritance is referred to as criss-cross inheritance.

***Crocus* spp.** The saffron.

Crop improvement and polyploidy At least one third of the angiosperms are polyploids. The proportion of polyploids among crop plant is even greater than this. Hence, it was thought artificial induction of polyploids might give very valuable result. Since the discovery of colchicine for induction of polyploidy in plants, several attempts were therefore made to evolve new and superior crop plants.

Crossing over A process in which genes are exchanged between non-sister chromatids of homologous chromosomes. Formation of chiasmata is the visible evidence of crossing over.

Crossover value The sum of the two recombination classes in a test cross expressed as a percentage of the total offspring.

Crotalaria juncea Fam. Papilionaceae Sunn hemp. A useful plant used as green manure and fibre.

***Croton* sp.** Garden croton bears many types of variegated leaves and can be propagated by cuttings.

Croton sparcliflorus Dichomatous branching. A weed.

Cruciferae Mustard family. Floral formula: ⊕, ♂, $K_{2+2}\,C_{4\,x}\,A_{+2+d}G\alpha$. A family of angiosperms to which many crops and vegetable belong.

Cryptogams These are plants that do not bear flowers or seeds and hence are commonly known as flowerless plants or seedless plants. One of the two subkingdoms of the plant kingdom and these were included in such category of plants in which sexual reproduction was not clearly seen.

Crystalloid Each aleurone grain is a solid, ovate or rounded body and encloses in it a large crystal like body known as crystalloid.

Crystals (mineral) The common forms of crystals found in plants are silica, calcium carbonate and calcium oxalate.

***Cucumis* spp.** Fam. Cucurbitaceae. *Cucumis sativus* (cucumber), *C. melo* is melon. Of both the fruits are important vegetable and fruits.

Cucurbitaceae A family of angiosperms. Floral formula is:

$$O\,K_{(5)}\,C_{(5)}\,A_3 \text{ or } 5$$

***Cucurbita* sp.** Fam. Cucurbitaceae. *Cucurbita pepo* (snake gourd), *C. moschata* are important vegetable plants.

Culm The jointed stems with solid nodes and hollow internodes.

Cumin Fam. Umbelliferae. *Cuminum cyminum.* It is a condiment.

Cumulative factors Non-allelic genes affecting the same character and enhancing each other's effect.

Curcuma amada The mango ginger.

Curcuma domestica The turmeric.

Curry leaf plant Fam. Rutaceae. *Murrya paniculata.*

Cuscuta Dodder plant. Fam. Convolvulaceae. *Cuscuta reflexa.* It is living plant parasite.

Custard apple *Annona squamosa.* Fam. Annonaceae.

Cuticle The outer walls of the epidermis are often thickened and cutinized. This cutinized layer check evaporation of water from the surface.

Cutin It is a waxy substance. It forms a definite thick or thin layer, called cuticle, on the skin of leaf or stem.

Cyathium A special kind of inflorescence found in Euphorbiaceae family. If has cup shaped involucre provided with nectatries.

Cycad One of the orders of gymnosperms.

Cycas it is evergreen palm-like genus mainly distributed in the eastern hemisphere. The genus has about 17 species out of which 5 species have been reported from India. It is a Gymnospermic plant.

Cyclic electron transfer In photosynthesis, one of the carriers possibly the first one appear to be ferredoxin. It is an iron compound containing coenzyme normally present in the chlorophyll. After ferredoxin, the high energy electron passes through a series of other carriers.

Cyclosis These are the streaming movements of the protoplasm enclosed by a cell wall and are performed around one or more vacuoles. These movements are of two types–(1) Rotation, and (2) Circulation.

***Cymbopogan* sp.** Fam. Gramineae. The lemon grass and of medicinal value.

Cymose It is an inflorescence in which the main axis ends in a flower and similarly the lateral axis also ends in a flower. Thus the growth of each axis is checked. The terminal flower is always older and opens earlier than the later ones.

Cynodon dactylon Fam. Gramine. A perennial runner grass. Important for making lawns and is good animal fodder.

Cypsela This is a dry fruit, developing on one side from an inferior bicarpellary ovary.

Cyst formation Cysts are formed by the cells of the genus *Azotobacter* where the entire bacterial cell rounds up and up and becomes surrounded by a thick wall. Also in the nematodes, it is seen that during their inactive season they engulf themselves in cysts. *See* Plant Pathology.

Cystolith In the leaf, sometimes crystals are deposited on a sort of stock which is the in growth of the inner epidermal wall. Finally whole crystalline mass looks like a bunch of grapes suspended from a stalk–this is cystoligh.

Cytocentrum The central apparatus. A structure formed by the centrosome together with the surrounding cytoplasm.

Cytochrome system In the aerobic respiration, 95 per cent ATP molecules are formed as a result of electron transfer from various hydrogen acceptors. These are NAD, NADP, FAD etc. The electrons are of received as a result of oxidation of Kreb's cycle and are high energy level. They are transferred from on hydrogen acceptor to the other in a definite sequence, their energy is stepped down and ATP molecules are formed from ADP. This systematic transfer of electrons occurs in inner layers of mitochondria and this whole system is known as cytochrome system.

Cytogenetics Study of the cellular structures and mechanisms associated with genetics. It is the interpretation of the phenomena of heredity in terms of cell structures.

Cytogenetic map A map showing the actual position of genes within the physical chromosomes.

Cytokinins The division of cytoplasm during cell division. Science discovered that kinetin (6-fufureldemine) induces cell division. The class of factors inducing cell division was subsequently called cytokinins.

Cytology The study of structure and function of the cell. It is a specialised form of detailed study of the cell, its contents and its behaviour.

Cytoplasm The portion of the cellular protoplasm which occurs between the plasma membrane and nuclear membrane. It is the protoplasm of the cell excluding nucleus, its outermost layer next to the cell wall is differentiated into a delicate membrane called plasma membrane, cell membrane, plasmalema or ectoplast.

Cytoplasmic male sterility Most of all of the pollen grains of such male sterile plants are abortive. This character is transmitted only through the female and never by the pollen. When all of the chromosomes of the male sterile line were replaced with chromosome of normal plants, the line still remained male sterile, showing thereby that male sterility is controlled by some agency in the cytoplasm.

Cytoplasmic sterility Plants carrying particular types of cytoplasm are male sterile but will produce seed if pollinated by pollen from male fertile plants. These seeds produce only male sterile plants since their cytoplasm is derived entirely from the female gametes.

Cytoplasmic movement The protoplasm of the cell performs various types of movements. Ciliary movement in which some free minute protoplasmic bodies such as zoospores and antherozoids of mosses and ferns posses special whip like or thread like structures known as cilia or flagella on their bodies. The amoeboid movements are the creeping movements of naked mass of protoplasm such as slime fungi and gametes in *Spirogyra*.

Cytoplasm's physical nature The cytoplasms consists of a semi-fluid ground substance of colloidal nature which is in sol-gel state at different times and in different cellular regions.

Cytosine A primidine base occurring in DNA and RNA. Pairs with guanine in DNA. ($C_4H_5ON_3$) occurring as a fundamental unit or base of nuclei acid. Two of the bases found in DNA are–(1) thymine (pyrimydine), and (2) cytocine.

Cytoskeleton A complex network of protein filaments interspersed throughout the cytoplasm.

Cytosol The parts of cytoplasm that occupies the space between founded organelles.

Cytotaxix Attraction of motile cells by specific diffusable stimuli emitted by other cells.

Daffodil Fam. Amaryllidaceae. *Naricissus* sp.

Dagger plant Fam. Liliaceae, *Yucca gloriosa.* Also known as Adam's needle.

Dahlia Fam. Compositeae. A garden ornamental for flowers of various hues.

Dalbergia sisoo Fam. Leguminoseae. Red wood tree. An important timber tree.

Dark reaction Blackman in 1905 proved that two types of reactions take place in the process of photosynthesis. In one case, the reactions can take place only in the presence of light. This is known as light reaction. In the second part, the reactions are independent of light and can take in dark also. This is called dark reaction.

Darwin In 1858, Charles Darwin (1809-1882) put forth the 'theory of natural selection'.

Date palm Fam. Palmae. *Phoenix sylvestris.* An important fruit bearing tree.

Datura metal Fam. Solanaceae. Thorn apple. Has medicinal value.

Datura stramonium It was for the first time in this plant that haploidy was recorded by Blakaslee. Later he noticed some trisomics in this plant.

Daucus carota Fam. Umbellifereae, carrot. An important vegetable.

Daughter chromosome A half chromosome during anaphase.

Davenport The author of the 'dominance hypothesis' postulated that a large number of favourable dominant genes contribute to increased vigour.

Deciduous If the leaf of any plant lasts one season, usually falling off in winter, it is called a deciduous plant/tree/leaf.

Decliny Unisexual or dicholinous flowers *i.e.* separate male and female flowers, may be borne by one and the same plant.

Decompound leaf When the leaf is more than thrice pinnate.

Decumbent When the stem is much branched and the branches spread out on the ground on all directions.

Decussate In opposite phyllotaxy on a pair of leaves is most commonly seen to stand at a right angle to the next upper or lower pair. Such an arrangements of leaves is decussate.

Defence mechanisms These are special organs developed by plants as arms of defence or have some special devices to repulse or avoid the attack of their enemies.

Deficiency The loss of a chromosome segment, especially a terminal segment, thus involving one, two or more genes.

Deficient chromosome If the genes of a normal chromosome are represented as ABCDEF and if two genes D and E are lost, the deficient chromosome will be ABCF.

Dehiscence When the pericarp of fruits burst to liberate the seeds, this is called dehiscence.

Delayed Not at the right time but later.

Deletion The loss of a chromosome segment, especially an interstitial segment.

Delonix regia Fam. Caesalpineae. Gold mohur. An ornamental avenue tree.

***Delphinium* sp.** Fam. Ranunculaceae. Larkspur. An ornamental herb.

Dentate When the teeth of the margin are directed outwards at right angles to the margin of the leaf.

Dentella repens Fam. Rubiaceae. Growing as weed.

Deoxyribose-nucleic acid (DNA) A usual double stranded helically coiled, nucleic acid molecule, compound of phosphate-dioxyribose 'back bones' of which are connected by paired nitrogen bases (purines and pyrimidines) attached to the dioxyribose sugar, the genetic material of all living organisms and many viruses.

Deplasmolysis Plasmolysed cells can recover their turgor if they are put again in pure water or in a solution with a lower osmotic concentration (hypertonic solution) than that of the cell contents. This phenomenon is called desplasmolysis.

Dermatogen This is the single outermost layer of cells of a stem. It passes right over the apex and continues downwards as a single layer.

Designs In any plant breeding experiment, proper statistically sound methods are applied while laying out the experiments in field.

Desmodium gyrans Fam. Leguminosae. Telegraph plant.

Desmosome Thickened areas of plasma membrane of two adjacent cells from which radiates out fine tonofibrils and which provide cellular adhesive to the cell.

Despiralisation In prophase stage, the chromosomes become shorter and thicker by a process (spiralization) that transforms a long thin chromosome into short thick coiled structure in the same way as a thin wire is converted into a coiled spring. As prophase proceeds, coils decrease in number and inccrease in diameter (despiralisation) and the chromosomes become increasingly more stainable, shorter and thicker.

Desynapsis The falling apart of the synapsing chromosomes before metaphase.

Detached method of grafting In this method, the scion branch is separated from the mother plant to be placed on the rootstock.

Detasseling Removal of male inflorescence (tassel) in maize.

Deterioration of varieties Some of the causes are (1) minor genetic variation (2) due to mutations (3) natural crossing (4) evolution of new races of disease causing organism deteriorate (5) developmental variations and (6) mechanical mixtures.

Development In growth of plants, development refers to the changing shape, form, degree of differentiation and the state of complexity of the organism.

Developmental variations Such variations are due to environmental changes.

Devil nettle Fam. Apocynaceae. *Alstomia scholaris*

de Vries (1848-1935) He introduced term 'mutation' from the large discontinuous variations (changes) in the genotype and proposed as 'Mutation Theory.'

Dew point The temperature of air at which it becomes saturated with water vapour, below the point water vapour starts to condense to form water droplets. It can be determined by using a hygrometer with tables based on dry and wet bulb thermometer or by using a special dew point hygrometer with a polished metal surface that can be cooled until a film of moisture appears at the temperature that is recorded.

Diadelphous When the filaments are united into two bundles.

Diaheliotropism Movement of plant leaves which follow the sun in such a way that they remain perpendicular to the sun's rays throughout the day.

Diakinesis A stage of prophase-I of meiosis in which paired chromosomes are much shortened and thickened. This is last stage in cell division in prophase of M-1.

***Dianthus* sp.** Pink.

Diaphorase Mitochondrial flavoprotein enzymes which catalyse the reaction dyes.

Diarch A root with two xylem bundles is known as diarch, with three as triarch, with four tetrarch, with five as pentarch and with more than five as polyarch.

Diastema Modified cytoplasm of the equatorial plane prior to cell division.

Dicentric With two centromeres.

Dicentric chromatid If in case of a chromatid inversion does not include the centromere, *i.e.* if it is a paracentric inversion, the meiotic anaphase will contain a chromatid with two centromeres and this is the dicentric chromatid.

Dichasial cyme In this type of inflorescence, the main axis ends in a flower and at the same time it produces two lateral younger flowers, sessile or stalked.

Dichogamy In many bisexual flowers, the anther and the stigma mature at different times. This condition is called dichogamy.

Dichogamous When the terminal bud of a stem bifurcates, producing two branches in a forked manner, this type of branching is dichotomous.

Dicot root in a circular section of such root, it shows three main regions (1) the outermost single layered epidermis or epiblema or piliferous layer, (2) the cortex and (3) the stele.

Dicotyledons Seeds bearing two cotyledons in their embryo.

Dictysomes The golgi complex which do not have secretary glands.

Didynamous If in a flower there are four stamens, of which two are long and two short, such stamens are called didynamous.

Differentiation Modification of different parts of the body for particular function during development of the organisms involving continual restriction of certain transcriptors.

Differentiation in buds A particular low temperature is essential for the dormant buds to vary in formation of vegetative or flower buds.

Diffusion Migration of molecules or ions, as a result of their own random movement from a region of higher concentration to the region of lower concentration. It requires no energy. The spread of molecules from a place where they are more abundant per unit of volume (and therefore collisions are more frequent) to places where they are less abundant per unit volume (where collisions are less frequent) is called diffusion.

Digitate In a palmately compound leaf, leaflets are commonly 5 or more, this condition is called digitate.

Dihybrid A cross between parents differing in two genes; or a heterozygote individual for two pairs of alleles.

Dihybrid ratio The crosses where two characters are involved.

Dill Fam. Umbelliferae. *Anetham graveolens* (sowa). A condiment.

Dimorphic In Gymnosperms, the plant is provided with two types of leaves–brown scale leaves and the spiral or foliage leaves alternates with the spiral of scale leaves. This condition is dimorphism.

Dimorphism In Alga *Volvox,* the ontogeny (development of an individual) of the colony reveals that all the cells similar during the early stages. However, a dimorphism becomes apparent due to the formation of large reproductive cells in the posterior hemisphere and small vegetative cells in the posterior hemisphere and small vegetative cells in the anterior hemisphere.

Dioecious individuals producing either sperm or egg (male or female) but not both. In dioecious, species sexes are separate. When unisexual or declinous flowers are borne on separate plants.

***Dionacea* sp.** A carnivorous plant. The Venus fly trap.

Dioscorea bulbifera Wild yam, produces bulbis in the leaf axils.

Diplobiontic In fungi (yeasts), the ascospores of the opposite type fuse and form a diploid cell. This diploid cell germinates and gives rise to diploid somatic cells. This type of life cycle is known as diplobiontic.

Diplococcus The cocci bacteria in which the cells are often joined in pairs.

Diploid An individual or cell having two complete sets of chromosomes.

Diplotene A stage of prophase-I of meiosis in which each of the synaptic chromosomes gets doubled by splitting.

***Dipterocarpous* sp.** The wood oil tree.

Discontinuous variation A mutation or sudden heritable variation.

Discordant Twins among which one member shows a genetic trait and the other does not show.

Disease That makes a plant or animal unhealthy or even kills them.

Disjunction The separation of homologous chromosomes during anaphase-I of meiosis.

Dismorphic heterostyly There are plants which bear flowers of two different forms one form bears long stamens and short style and other form may bear long style and short stamens. This condition is termed as dimorphic heterostyly.

Disome An organism with two chromosomes of each kind.

Distant hybridization Hybridization between individuals belonging to two different species of the same genus or of the different genera.

Distichous This also called phyllotaxy 1/2. In this condition, the third leaf stands over the first, thus there are two orthostitches *i.e.* two rows of leaves and therefore the phyllotaxy is distichous.

Distyly In Primula, there are two types of flowers borne on different plants: (1) thrum with short styles and long filament; (2) pin with long styles and short filaments. Pollinations are compatible only between stigmas and anthers of the same height *i.e.* between thrum and pin and the reciprocal. This state of styles present in a species is called distyly.

DNA Same as Deoxyribonucleic acid.

Dolichos lablab Fam. Papilionaceae. Country bean. Important vegetable plant of which pods are used.

Dominance Intra-allelic interaction so that one allele manifests itself more or less when heterozygous than its alternative allele.

Dominance hypothesis One theory of heterosis is based on dominance, on the assumption that hybrid vigour is entirely due to the bringing together in the first generation hybrid of a large number of favourable dominant genes contributing to vigour. It also postulates that genes that are favourable for growth and vigour are dominant, and genes harmful to the individual are recessive.

Dominant character A character possessed by one parent of a hybrid which appears in the F_1 to the exclusion of the allelic character from the other parent.

Dominant mutation Some mutations are dominant but most mutations are recessive. A dominant mutation can be detected immediately after its occurrence but a recessive mutation may not be found easily especially in cross fertilised plants where they may be carried along in heterozygous condition for several generations.

Donor parent That parent from which by back crossing, one or more genes are transferred to the back cross plant.

Donor sources When genes are identified to have come from different locations in the country or in the world. These genes (characters) are induced in a crossing programme for its improvement.

Dormant Very young or inactive stage of plants.

Dorsiventral leaf When the leaf is flat with the blade placed horizontally, showing distinct upper surface. Such leaves are strongly illuminated on the upper surface then on the lower. This unequal illumination induces a difference in the internal structure between the upper and the lower sides.

Double cross A cross between two F_1 hybrids.

Double fertilization The phenomenon–two fusions (in angiosperms) one between the egg and one of the male gametes (true fertilisation) and the other between the second male gamete and secondary nucleus, is known as double fertilization.

Dragon plant Fam. Liliaceae. Dracaena sp.

***Drocera* sp.** The sundew plant. It is a carnivorous plant.

Drosophilla Thomas Hunt Morgan found that some 'factors' in the fruit fly *Drosophila melanogaster* were not inherited at random, but in groups. This fly was later extensively used in genetical and cytological studies for its having few chromosomes and quick reproduction.

Drought resistance Draught is the deficiency of soil moisture available to the plant so that it checks all plant activities. Although primarily the draught injury is due to deficiency of soil moisture, its adverse effects are aggravated by atmospheric factors such as high temperatures, low humidity and wind etc.

Drumstick *Moringa* sp.

Drupe This is fleshy, one or more seeded fruit with the peri carp differentiated into outer skin or epicarp, often fleshy or sometimes fibrous mesocarp, and hard and stony endocarp and hence this fruit is called stone-fruit.

Drying This is an important process for preservation of seeds and horticultural products etc.

Dryland farming Same as rainfed farming.

Dryopteris This is a fern, a big group of highly advanced cryptogams. It is a member of *Pteridium* commonly as broken ferns. This genus is having about 150 species and is found in all regions of the world, but is particularly rich in subtropical and warm temperate regions.

Duckweed *Lemna* sp. It is floating plant on the water surfaces in the lakes and other such water bodies.

Duplicate genes Identical genes but located at different loci and showing no cumulative effect.

Duplicate ratio Crosses involving duplicate genes give the phenotypes in the ratio 15: 1 in the F_1.

Duplication The occurrence of a segment twice in the same chromosome or in the same complement.

Duramen In old trees, the greater part of secondary wood is filled up with tannins, resins, gums, essential oils, etc. which make it hard and durable. This region is known as heartwood or duramen. The central part of the dead wood which is darker in colour than the surrounding functional xylem is known as heart wood or duramen.

Duranta This plant is used as a hedge plant and can be propagated by cuttings.

Dwarfism To reduce in size.

Dwarf male In alga (Oodogonium), the androspore swims for a while and soon attached itself direct to the oogonism or to a cell close to it. It then produces a short narrow filament, called the dwarf male.

Dwarf shoots In Gymnosperms *(Pinus)*, these are the branches of limited growth borne on the long shoots and are commonly known as dwarf shoots or foliar spurs or spurs.

Dyad At telephase-1, a nuclear membrane may be formed around the group of chromosomes at each pole. In plants, a cell wall is formed at the equator and the two cells together are known as a 'dyad'. A pair of homologous chromosomes before their duplication in meiosis.

Echinops **sp.** The globe thistle. This is a desert plant in which the leaf surface is too much reduced to protect it from much transpiration.

Eclipta alba Fam. Compositeae. A common weed.

Ecological niche The role played by a plant or animal in the functioning of an ecosystem *e.g.* the ecological niche of a pike is as a fresh-water predator of other fish. An organism fulfilling its niche exerts influence on its neighbours in the community and its abiotic environment.

Ecology The scientific study of the relationship of a plant or animal to its natural environment. This includes the organism's responses to its physical surroundings (*i.e.* to weather or the changing of day length) as well as its interaction with other organisms (such as who eats whom). Applied ecology seeks to improve man's use of environmental resources by translating lessons learnt from ecological research to such problems as the populations control of pests.

Economic plants These plants have variety of uses. Many of them occur in nature, particularly on hills and forests, while a good number of them are cultivated for food and industry.

Ecosystems *See* Soil Science.

Ectogony The influence of pollination and fertilisation on structures outside the embryo and endosperm. Effect may be on colour, chemical composition, ripening or abscission.

Ectophyte A plant which lives externally on another organism.

Ectopic pairing Pairing between non-homologous segments of salivary glands' chromosomes in Drosophila, presumably involving mainly heterochromatic regions.

Ectoplasm The outer gelled zone of the cytoplasmic ground substance in many cells. It is a part of cytoplasm–its outer surface forms an extremely thin and delicate membrane called the plasma membrane or ectoplasm.

Edaphon Flora and fauna inside the soil.

Egg The female gemete. In angiosperms, during the development of female gametophyte, four of the nuclei lie at the end of the embryo sac and the remaining four at the other end. One nucleus from each pole moves towards the centre and is known as the polar nucleus. These two polar nuclei ultimately fuse giving rise to a diploid nucleus known as secondary nucleus or fusion nucleus. The remaining three nuclei at each end of the embryo sac organise themselves into cells. The three cells at the chalazal end are known as the antipodal cells and three cells towards micropylar end constitute the egg apparatus. The central in the egg apparatus is female gamete and is known egg or ovum or oosphere. The other two cells are the synergids or help cells.

Egg apparatus In the mature-sac, a group of three cells (in angiosperms) each surrounded by a very thin wall may be seen always lying towards the micropyle; this group is called the egg apparatus. One cell of this group is the female gamete (reproductive unit) known as the egg cell or ovum and the other two are known as synergids.

Egg nucleus Fertilisation consists essentially of the fusion of the egg nucleus of the embryo sac and the sperms nucleus of the pollen grains, resulting in the formation of the zygote, the first cell of the new individual.

***Eichornia* sp.** The water hyacinth. An aquatic plant.

Einkorn The species of Triticum can be grouped according to the number of chromosomes into the diploid Einkorn group (2n =

14), the tetraploid Emmer group (2n = 28) and the hexaploid Vulgare group (2n = 42).

Elators In Bryophyta, the oospore germinates in situ and gives rise to the sporophyte which reproduces asexually by spores. The sporophyte is a complex body and is known as the sporangium. It consists of a food, a short stalk called seta and a capsule. The capsule stalk consists of a single layered wall and a mass of small cell. Some of these cells grow up into elongated, spindle-shaped, spirally thickened structures called elaters.

Electrophorasis The migration of suspended particles in an electric field.

Electron microscope It can magnify as high as 2,00,000 times (diameters) of even more. It was invented by two German scientists. Knooll and Ruska in 1932.

Electrons When compounds absorb energy from non-ionising radiations, their electrons are raised to higher energy levels.

Elephant ear plant A few adventitious buds are produced on the surface of the leaf from the veins and also from the petiole.

Eleusine coracana Fam. Gramineae. A smaller millet.

Emasculation Removal of anther from the flowers.

Embryo The entire fleshy body, as seen after removing the seed coat. As the seed germinates, it gives rise to seedling which gradually develops into a full plant. In pteridophyta, the oospore starts growing and gives rise to an embryo. *See* embryo sac.

Embryo sac The female gametophyte arising from the megaspore by meiotic division.

Embryo sac mother cell A cell from which by meiotic division develop four megaspores, one of which develops into the embryo sac. Abbreviated as E.M.C.

Emmer *See* Einkorn.

Endarch Xylem is endarch in stems and its development is centrifugal. The first formed xylem or protoxylem consists of annular, spiral or sclariform vessels; it lies towards the centre of stem and its vessels have small cavities. As far as the xylem is concerned, protoxylem lies towards the centre of the axis and the metaxylem towards the periphery. Such a position of the protoxylem or bundle is known as endarch.

Endiva Fam. Compositeae. *Cichorium endivia*. Medicinal.

Endocarp The pericarp of a fruit may be thick or thin; when thick, it may consist of two or three parts: the outer called epicarp, forms the skin of the fruit; the middle, called mesocarp is pulpy and the inner is called endocarp. In fleshy fruits, the fruit wall is fleshy at maturity and is distinguished in three layers–the outermost exocarp, or epicarp, the middle mesocarp and the innermost endocarp.

Endodermis it is the single internal layer of the cortex often wavy, also called starch sheath as it often contains numerous starch grains. It is the innermost layer of the cortex and is composed of vertically elongated closely packed cells which appear rectangular or barrel shaped in cross section.

Endogamy Pollination of a flower by another flower of the same plant.

Endogenate The part of the bacterial chromosome that is homologous to a genome fragment (exogenome) transferred from the donor to the recipient cell in the formation of merozygote.

Endogenous Lateral roots originate from an inner layer; so they are said to be endogenous. The development of branch root from the inner layer *i.e.* pericycle, is described as endogenous in contrast to the exogenous origin of stem branches.

Endomycetales Yeast is a common name given to more than one genera belonging to family Saccharomycetaceae of the order Endomycetales, which in turn belong to sub class hemiascomycetes.

Endoplasmic reticulum A tubular double membrane system in the cytoplasm, continuous with the nuclear membrane and bearing many ribosomes. Containing many enzymes and have skeletal, transporting and biosynthetic functions of the cytoplasm.

Endosperm In plants, one of the sperm nuclei fuses with the egg to form the zygote and the other sperm nucleus fuses with the fusion nucleus to form endosperm.

Endosperm nucleus The second male moves towards the centre of the embryo sac fuses with the secondary nucleus to form a triploid nucleus known as primary endosperm nucleus.

Endospore When the conditions are unfavourable in bacteria, the cell contents including the genetic material shrink into a spherical or oval mass which becomes enclosed in a single or double layered protective wall forming an endospore.

Endosporium In algae, the zygospore is shining dark brown in colour ellipsoidal in outline. Its wall is circularised and is differentiated into three layers–an outer thick and hard exposporium, a middle mesosporium and an inner thin and delicate endosporium. In fungi (Mucor), the fusion of the nuclei results into the formation coenozygote which secretes a thick waxy wall around itself. This wall is differentiated into two layers–an outer thick rough and dark coloured exosporium and an inner thin endosporium. The coenozygote is now known as zygospore. *See* also for Bryophyta.

Energy Capacity for doing work.

Enhydra fluctuans Fam. Compositeae. Has medicinal value.

Enryplastic organism An organism with a marked ability to change and adapt a wide specturm of environmental conditions.

Entida gigas Fam. Mimoseae. The nicker bean.

Entomophily Pollination by insects.

Environment The sum total of the external conditions which affect the growth and development of an organism.

Environmental variation A non heritable variation caused by the reaction of the individual to an environmental change.

Enzyme Any substance, protein or in part that regulates the rate of any biochemical reaction in living organisms.

Epibasal In a pteridophyta embryo, the basal cell which divides the oospore inot two halves–an interior or the epibasal lying towards the knotch of the prothallus and posterior or the hypobasal lying towards the posterior end.

Epiblema The dermatogen continues upwards as a single outermost layer called the epiblem. The outermost layer of cells in the case of root is most often from the outermost layer of ground meristem, the protoderm having been worn out by friction with the soil particles and is known as epiblema.

Epicalyx This is a whorl of nacteoles developing at the base of the calyx.

Epicarp *See* endocarp.

Epicotyl During germination, the cotyledons remain under the soil.

Epidermis It is the single outermost layer, which extends over the entire surface of the plant body. It is the jacket or layer of cells around the plant and is present in all young organs. Sometimes the epidermis is made up of many layers of cells and is called multiple epidermis.

Epigeal In some seeds at the time of germinating, the cotyledons are seen above the soil surface.

Epigyny In such flowers where thalamus grows further upward, completely enclosing the ovary and getting fused with it, and bears the sepals, petals and stamens on the top of the ovary.

Epinasty When in a flowering plant growth is more rapid on the upper surface.

Epipetalous As per adhesion of the stamens, when the stamens adhere to the corolla wholly or partially by their filaments, anthers remaining free, they are said to be epipetalous.

Epiphytes Such plants that grow attached to the stem or branches of other plants but do not suck them.

Epiplasm The entire protoplasm of an ascus is not used up in the formation of ascospores but there is left a balance of protoplasm called epiplasm.

Episomes In bacteria, *E. coli*, the maleness is determined by the presence of extrachromosomal genetic particles–the episomes.

Epistasis A kind of interaction between genes belonging to different pairs of alleles, the dominant allele in one of the pairs preventing the dominant allele in the other pair from expressing itself. Thus the gene A maybe epistatic over B. B is then said to be hypostatic under A.

Epistatic ratio The 12 : 3 : 1 dihybrid ratio. When two non-allelic genes affect the main part or trait of the organisms, it is likely that the expression of one covers up or hides the expression of the other.

Epithem Some cells of the mesophyll are without chloroplast and are loosely arranged having large intercellular space and are termed as epithem.

Epithelium In plants like most cereals, the cotyledon is known as the scutellum. It supplies the growing embryo with food material absorbed from the endosperm with the help of the epithelium.

Equational division In the higher plants, meiosis takes place in the ovaries and the anthers. Meiosis comprises of two divisions of the nucleus but one division of the chromosomes. The first division is called reductional division (or meiosis I) because a reduction in chromosome number occurs during this division and the second division is called equational division.

Equator At metaphase, a large number of delicate fibres, oriented roughly parallel to one another. The extreme ends of the spindle are called the poles and the central region, the equator.

Equatorial plate The figure formed at the spindle equator in nuclear division.

Equatorial rainforests There is wide diversity of plant species and associated fauna and a lack of seasonality in flowering and fruiting. Such vegetation occupies equatorial lowland in South America, Central America and SE Asia. These formers produce valuable hard woods such as mahogany but this resource is much endangered by over exploitation.

Equisetum Fam. Equisetinae. Horsetail.

Ervatamia coromaria Fam. Apocynaceae. It is an ornamental hedge plant.

Erythrina variegata Fam. Leguminoseae. Coral tree.

Escherichia coli This bacterium has been a material for studies responsible for this rapid progress made in recent years in the fields of physiological and biological genetics. Other organisms made use of for such studies are bread mould Ineurosporea and the typhoid bacterium Salmonella.

Estimation of linkage from F_2 data *See* calculation of linkage by additive method.

Etario An aggregate of simple fruits borne by a sing flower.

Etiliotated Without light chlorophyll cannot develop; continued absence of light decomposes chlorophyll and the plants become etiolated.

Etioblast An immature chloroplast containing prolamellar bodies.

Etiolation When plants are kept in darkness, they lead to pathological condition known as etiolation.

Eupatorium odoratum Fam. Compositeae.

***Euphorbia* spp.** Fam. Euphorbiaceae. Serious weeds of cultivated crop fields.

Euploid It is an individual containing in its somatic cells more than two sets of chromosomes and whose somatic number of chromosomes is an exact multiple of the monoploid (or haploid or basic) number. Also can be called a polyploid.

Euploidy Variation in chromosome number by whole st or exact multiple of the monoploid (haploid) number *e.g.* diploid, triploid. Euploids above the diploid level may be referred to collectively as polyploids.

Evaporimeter An instrument used to measure the amount and rate of evaporation of a liquid. Also called atmometer. There are two main types–one measures the loss in weight of a known quantity of liquid and another measures the change in level of surface of the liquid.

Evening primorse *Oenothera lamarkiana.* This plant has been put to extensive studies on mutations especially by de Vries.

Evolution Hereditary changes lead to evolution in plants. The problems of origin of different forms of life are discussed under this branch of biology.

Exalbuminous *See* albuminous.

Exarch Xylem is exarch in roots and its development is centripetal. In the case of roots of flowering plants, the differentiation of xylem elements starts from the outside of the procambium and proceeds inwards with the result that protoxylem lies towards the periphery. Such a xylem bundle is said to be exarch.

Exine Each pollen grain consists of a single microscopic cell and possess two coats: the exine and the intine. The exine is tough cutinized layer which is often provided with spinous outgrowth or markings of different patterns sometime smooth. The intine is, however, a thin, delicate cellulose layer lying internal to the exine.

Exodermis A few layers internal to the epiblema representing the outer zone of the cortex.

Exoenzymes The bacteria which derive their food from the dead organic matter are named as saprophytic. They liberate hydrolytic exoenzymes into their surrounding medium to convert the food stuffs into diffusable form.

Exogenous The lateral roots originating from the few outer layers.

Exogenous variations Same as developmental variations.

Exogenote Chromosomal fragment homologous to an endogenote and donated to a merozygote.

Exosporium *See* endosporium.

Exotic A plant or animal that is of foreign origin and not native to the area it is found.

Exponential growth The situation in which a variable increases by addition (or decreases by subtraction) and the outer variable increases by multiplication (or decreases by division). The line graph that expresses this relationship is known as the exponential curve or J-curve. When data with exponential relationship is plotted on semilogarithemic paper, the curve is shown as a straight line.

Extrapolation Estimating the value of a quantity or function outside the known range of values. For example, a linear graph showing population growth can be extended beyond most recent values on the graph. The further from the known range the line is taken the greater is the risk of error in the extrapolation.

Exudation The excess water is got rid of in many herbaceous plants and undershrubs by a process commonly known as exudation or guttation. Accordingly, water escapes in liquid form at night and accumulates in drops at the end of veins. Exudation takes place in the absence of transpiration.

Eye spot Also called stigma. In algae, the eyespot functions as an organ for determining the direction of movement. Its red colour is due to the presence of a pigment hetrochrome. *Ichlamydomonas pluristigma* the number of eye-spots may be more than one.

F_1 The first generation after a cross.

F_2 The second filial generation obtained by self fertilization or cross inter se of F_1 individuals.

F_3 Progeny obtained by self fertilizing F_2 individuals.

F_2 **breeding behaviour** In a dyhybrid cross, the F_2 seeds are genotypically of nine different kinds can be shown from their breeding behaviour.

Factor Gene, character.

Facultative anaerobes Majority of bacteria come in this category. They can grow best in oxygen and also can live as anaerobes.

Family It is a group of genera which shows general structural resemblances with one another, mainly in their floral organs.

Farmer's potometer It contains a wide mouthed jar provided with a tight fitting cork or rubber stopper having three holes. The cork is provided with dropping wax funnel, carbon thick capillary tube passing only to the lower level of the cork of the bottle. Behind the tube a scale marked in centimetres. Through the third whole is passed a shoot which has been cut under water. This apparatus is used to measure the rate of transpiration in a plant.

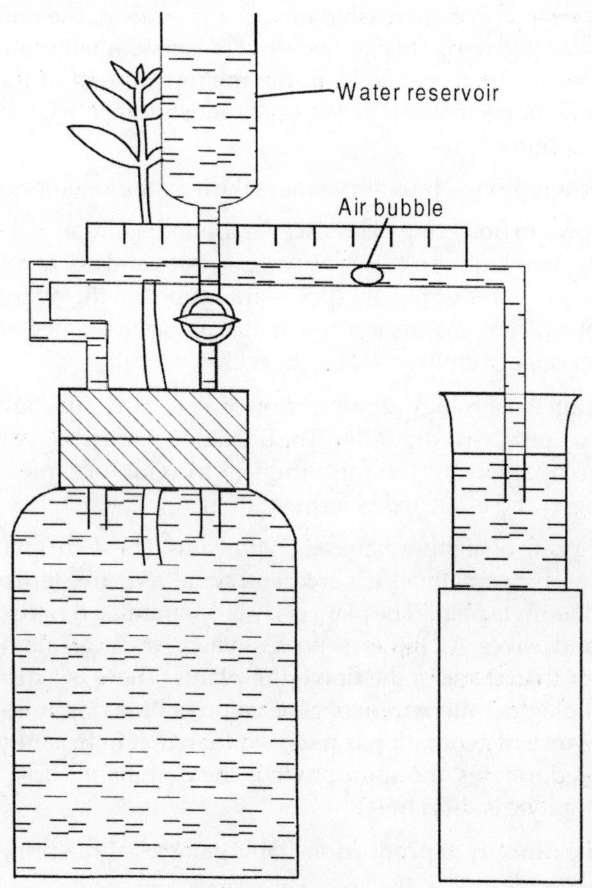

Farmers' Potometer

Fascicular cambium The cambium as is the left over of the apical meristem and is primary in origin. It is also known as intrafascicular cambium or fascicular cambium.

Fasciculed roots When several tubular roots occurs in a cluster or fascicle at the end of base of the stem.

Fats These are the glycerides of fatty acids.

Fecundity The innate reproductive capacity of the individual organisms, as denoted by its ability to form and separate from the body of the mature germ cells.

Female gamete Three successive mitotic divisions of the embryo sac nucleus give rise to eight haploid nuclei within the embryo sac. One of the three nuclei at the micropylar end of the sac becomes the egg or the female gamete and the other two become the synergids.

Female gametophyte The embryo sac is the female gemetophyte.

Female spores In flowering plants, we *See* around us are sporophyte bearing two kinds of spores *viz.* megaspores and microspores. The megaspores or female spores are formed in the ovaries of the flower. They are formed from megasporocytes or megaspore mother cells, of embryo sac mother cells.

Fermentation Anaerobic decomposition of an organic substance in the body of a living organism. The breakdown of carbohydrates is accompanied by the formation of alcohol, organics and hydrogen and is termed as fermentation.

Fern A big group of highly advanced cryptogams. The stem is mostly rhizome, but sometimes it is erect and aerial, as in tree ferns. It is a Pteridophyta plant. The plant body is differentiated into roots, stem and leaves. All the parts possess vascular tissues although simpler than those in the flowering plants. There is a distinct morphological alternation of generation in Pteredophytes but the balance of generation is reversed than that in Bryophytes. In Pteredophytes, the sporophyte is the dominant phase and gametophyte is short lived.

Fertility The capacity of production viable gametes as offspring.

Fertilization Fusion of the nuclei of the male and female gametes. Also *See* Soil Science.

Ferula asafoetida Fam. Umbellifereae. Asafoetida. Used as a condiment after extract the product from the roots of the plant. Used as a medicine also.

Fever nut Fam. *Caesalpinia bonducells.* Medicinal plant.

Fibres Commercial vegetables fibres are–(1) floss, fibres or lint (Cotton, *Gossypium* sp.), silk cotton (*Bomba* sp.), (2) Bast fibres; jute (*Chorchorus* sp.), sinhemp (*Gotalasia* spp.); and rhea or ramia (*Mohermeria* sp.), (3) coir fibre, coconut fibres and leaf fibres, (4) Bowrsing hemp (*Sanserieria* sp.) and American aloe (Agave). In their usefulness to man, fibre yielding plants are next to food plants. Fibres are mostly thick cells long wall containing legnin and cellulose. Their lumen is small and ends are pointed.

Fibrous root When a primary root does not persist, a cluster of slender roots is seen to grow from the base of the stem. These are the fibrous roots.

Ficus pumila Indian ivy. It has a stem climbing by the help of rootlets.

Fig Fam. Moraceae. Ficus glomerata.

Filament Each stamen is made up of three parts–filament, anther and connective. The stamen has a long or short stalk called filament.

Fimmonia acidissima Fam. Rutacene, Elephant apple.

First filial generation The first progeny after a cross (F_1).

First pollen mitosis This nucleus of the pollen grain divides by mitosis (first pollen mitosis) into two, one called the tube nucleus and the other called the generative nucleus.

Fisher Fisher R.A. developed the statistical methods which increased the efficiency of experiments.

Fission Bacteria commonly divide by the process of fission. A constriction appears around the middle of the cell and it becomes split up into two parts. Vegetatively or asexually yeasts may reproduce by fission as do the bacteria but majority of them reproduce by budding. These which reproduce by fission are called fission bacteria.

Fitness The number of offspring left by an individual as compared with the average of the population. Reproductive value determined by a genotype in a population.

Fixation Chance fluctuation of members of a pair of alleles around the mean that may result in homozygosity of one or the other allele and that henceforth will prevent segregation between the different alleles.

Fixed population A population in which allelic frequency of a given allele is 1 per cent.

Flagella Several types of bacteria are provided with one or more slender whip-like threads called flagella. Many bacteria perform locomotion with the help of small whip like appendages called flagella.

Flax Family Linacae. *Linum sativum.* A plant with pale blue flowers that is grown for the fibre obtained from its stem and for its seed, which when crushed produce linseed oil. Flax stalks are processed to provide the fibre that is used to make linen, fine writing paper and cigarette paper. It is a temperate climate crop.

Flea seed Fam. Scrophulariaceae. The common weed plant. *Plantago* sp.

Flemming In 1882, Flemming discovered mitosis (mitos-thread), because the chromosomes appear as threads in the early stages of cell division.

Floral diagram The number of parts of a flower, their general structure, arrangement, aestivation, adhesion, cohesion and position with respect to the mother axis are represented by this diagram.

Floral formula The different whorls of a flower, their number, cohesion, adhesion and their relative position may be represented by this formula.

Flower It is a highly modified shoot meant essentially for reproduction of the plant. A flower is a modified shoot meant for reproduction. A typical flower is made up of four sets of floral parts–sepals, petals, stamens and carpel or carpels.

Flower bud The dormant stage of flower.

Flowering plants These plants mainly consist of our economical plants, used as food, fruits, fibres, vegetables etc.

Fluctuating variations Same as developmental variations.

Fluorescence The chlorophyll solution absorbs through transmitted light appears deep green colour, but by reflected light it appears blood red. This property of chlorophyll is called fluorescence.

Foeniculum vulgare Fam. Umbelliferae. Anise or fennlel is an important condiment and medicinal plant.

Follicle A dry one hambered fruit dehisces by one suture only.

Food plants With respect to the economic importance of plants as a whole they are classified as food grains, fodder, sugar, building materials, for use in paper industry, perfumery, plants having aesthetic value etc.

Foot In Bryophyta foot is short, poorly developed, acuminate type and is embedded in the stem tip from where it draws nourishment for the growing sporogonium.

Foot cell in fungi *Aspergillus,* the vegetative cell from which a coniodophore is formed is called a foot cell.

Formative element The data from these even crosses led Mendel to suggest that the members of a pair of 'differentiating elements' in a hybrid actually separate themselves during the formation of gametes that half of the gametes carry one 'formative element' and the other half carry the other element.

Formee speciales The fungus *albugo* consists of a number of biological races of formae speciales, each of which (a formae specialis) is restricted to an individual host species.

Fossils The remains of ancient plant preserved in rocks are called fossils.

Foundation seed Seed produced from breeder's seed by the original breeder or an agricultural research station. It is the source of certified, either directly or through the registered seed.

Founder principle The premise that a new isolated population diverges from the parent population because of sampling error, parent and daughter populations on different gene pools.

Fragaria elatior Among plants *Fragaria elatior* is one in which the female ZW and male is ZZ type.

Fragaria vesca Fam. Rosaceae. Strawbery.

Fragment A chromatid without a centromere or acentric fragment.

Fragmentation In bacteria, the vegetative reproduction takes by binary fission or fragmentation or by budding.

Free cell formation This is a modification of indirect nuclear division. It differs from the latter in that the cell wall is not formed immediately after the division.

Free martin Twins of cattle, one of them being normal male and the other sterile female produced due to diffusion of male hormone with the female through the common placenta.

Free recombination Genetic recombination having a frequency of 50 per cent that is occurring by independent reassortment.

Fructification As for structure of fungi, the aerial portion of fungus constitutes its main body and is the 'fructification' or fruit body of the plant.

Fruit After fertilization, the ovary also begins to grow and gradually it matures into a fruit. It is a structure composed of one or more matured ovaries with or without seeds together with any accessory floral part.

Fuel Plants are also used as fuel. A good fuel is that which produces sufficient heat energy, burns slowly and is not smoky.

Funaria It is a large genus of Bryophytes having about 117 species which are distributed throughout the world.

Fungi The simplicity of structure, the absence of chlorophyll and storage of food other than starch, bindi all the groups together in a single group 'fungi' which have many characters in common.

Fungus It is Thallophyta without chlorophyll and most of them being microscopic organisms. Economically they are important group for human beings.

Funicle Each ovule is attached to the placenta by slender stalk known as funicle. The nucellus of an ovule is attached to the placenta by means of a short called funicle.

Fusiform root When the root is swollen in the middle and gradually tapering towards the apex and the base, being more or less spindle shaped in appearance.

Fusion nucleus One of the sperm nuclei fuses with the egg to form the zygote and the other sperm nucleus fuses with fusion nucleus to form the endosperm. The three nuclei at the chalazal ends of the embryo sac are called the antipodals. The two nuclei at the centre unite to form the fusion nuclei.

Galton It was observed that populations were found to exhibit a continuous range of variation in respect of measurable characters on which Francis Galton showed that this variation could be plotted on a graph as a 'normal curve'.

Gametangium In sexual reproduction of *Mucor* (fungi), each branch of the hyphae swells at the tip to become a club-shaped progametangium. Soon a transverse appears near the tip of the progametangium and divides it into a terminal pair gametangium.

Gamete A protoplast (sex cell) which in the process of sexual reproduction, fuses with another protoplast, one of them may be termed as male and another female.

Gametes The two sexual reproductive units.

Gametic purity A gamete or even a hybrid is pure in case it has either the dominant allele or the recessive allele but never both.

Gametic mutation A mutation occurring during gamete formation. It is a gene mutation in one member of a pair of homologous chromosomes, occurs in a cell of the 'germ line' before meiosis, half the number of gametes derived from the cell would bear the gameted genes. If the gene mutation occurs only in one gamete (*i.e.* after meiosis) the gamete is one that would have the mutant gene, if the mutation is a dominant one, the new character would be observed in the individuals produced by the mutated gametes.

Gametic number The number of chromosomes in the nuclei of the gamete.

Gametogenesis The origin, mutation and differentiation of gametes from the generative cells.

Gametophyte Looking to any higher cryptogam, one generation reproduces by asexual reproduction method *i.e.* by spores and the other by the sexual method *i.e.* by gametes. The former is therefore called sporophytic or asexual generation and the latter the gametophytic or sexual generation.

Gametophytic system The pollen tube reaction is determined by the genotype of the pollen tube (*i.e.* by the genotype of the gametophyte). Plants with gametophytic incompatibility, therefore, produce pollen with two types of behaviour *e.g.* plants of genotypes S_1 S_2 produce pollens of two types, one type behaving as S_1 allele and the other type behaving as S_2 allele. The somatic tissue of the style of a diploid plant has two different alleles *e.g.* S_2 S_3 and one of these alleles, say S_2 is similar to the S allele of the pollen tube *i.e.* S_2 the growth of the pollen tube is inhibited. The inhibition occurs at some stage during pollen tube growth in the style. Gametophytic incompatibility is found in *Prunus, Trifolium, Lycopersicum, Solanum, Nicotiana* etc.

Gamma rays These rays have shorter wave length than X-rays and hence are more penetrating. Important source of pure gamma rays of high energy is the radioactive isotope of Cobalt, Co^{60} readily available from atomic piles.

Gamopetalous Like the calyx, the corolla may be gamopetalous or polypetalous, according as the petals are united or free.

Gamophyllous In flowers, sometime the perianth whorl is usually free or polyphylous and sometime it may be united or gamophyllous.

Gamosepalous The sepals being united.

Ganong's potometer It consists of a horizontal graduated tube with one end bent downwards with a small air opening and the other bent upwards and having a wide end for fitting a twig.

This apparatus is used for measuring the rate/amount of transpired water.

Gardenia jasminoides Fam. Rubiaceae. Ornamental plant.

Garlic Fam. Liliaceae. *Allium sativum.* An important vegetable and medicinal plant.

Gas natural *See* Soil Science.

Gel Jelly-like state of protoplasm.

Geminiflorus A condition having flowers in pairs.

Gemmae cup In Marchantia, special multicellular bodies known as gammae, develop in gemma-cups on the thallus for the purpose of vegetative propagation.

Gemmules Darwin thought that hereditary particles termed pangenes or gemmules are produced by every part of the body during the lifetime of the organism and that, these assume the characters of the various parts of the body from which they were derived.

Gene The unit of inheritance which is located at fixed locus on the chromosome and which by interaction with other genes, the cytoplasm and the environment governs the development of a hereditary character. Gene is a small point on a chromosome having a unitary biochemical function and a specific effect on the properties of the individual.

Gene interaction The action on non-allelic genes upon each other.

Gene mutation A heritable variation due to an alteration of a definite locus on a chromosome.

Gene pool The available variability found in a plant species. Mango's gene pool is biggest in India.

Generation (of seed) S.G. 1 is seed directly produced by cross-pollination and from which F_1 plants will grow; S.G. 2 represents self bred seed from F_1; S.G. 3 is self bred seed from a F_2 etc.

Generative nucleus The pollen grain nucleus divided by mitosis (first pollen mitosis) into two; one called the tube nucleus and the other called the generative nucleus.

Gene symbols For genetic studies, it is customary to symbolise gene by letters. The dominant gene is represented by a capital letter and its recessive allele by the corresponding small letter.

Genetic emasculation Use of genetic factors to make the male gamete non-functional in self and/or cross pollination.

Genetic erosion Gradual disappearance of various forms of cultivated species and of wild relatives.

Genetic load The proportion by which the fitness of the genotype is decreased by deleterious genes as expressed in lethal equivalents of genetic death.

Genetic map Morgan postulated that genes are arranged in a linear order along the length of the chromosome, each gene having a fixed place on the chromosome and its allele a corresponding position on the homologous chromosome. In mapping genes, a unit distance must be used and this unit is called a map unit which is the space within which one percent of crossing over takes place.

Genetics The science dealing with heredity and variation. Genetics seeks to account for the resemblances and differences exhibited by organisms related by descent.

Genetic sterility Sterility of individuals in which there is completely normal pairing of chromosomes is attributed to genes because, shortly after meiosis, the embryo sacs and pollen grains deteriorate even though such racial hybrids show no disturbances in chromosome pairing. *See* oryza sativa.

Genetic variation A heritable variation produced by a change in the genes *e.g.* a mutation, a recombination etc.

Genome or genom A chromosomes' set *i.e.* the chromosome complement of gamete.

Genomic allopolyploids Containing two more genetically different chromosomes' sets from two or more species. These are also called alloploids.

Genotype The entire genetic constitution of an organism.

Genus it is a collection of species which bear a close resemblance to one another in morphological character of the floral or the reproductive parts.

Geotropism It is the movement of plant organs in respect to the force of gravity and has a marked effect on the direction of growth of plant organs.

Geotropism The gravitational force has a stimulating influence on the growth of a plant organs. The roots grow down towards the force of gravity and are positively geotropic while the shoots grow upwards against the force of gravity and are negatively geotropic. Dorsoventral organs such as leaves, runners, lateral roots tend to place themselves at right angle to the force of gravity and are termed as diageotropic. However, some special roots like aerial roots and pneumatophores are negatively geotropic.

Germ cell Gametes of cells which give rise to gametes.

Germination The process by which the dormant embryo wakes up and begins to grow to form a seedling.

Germplasm The material basis of heredity. The heredity substance in the germ cells.

Germplasm theory Weismann (1834–1914) a German zoologist, suggested in 1887 that a reduction in chromosome number took place during the formation of the egg and the sperm, and that the original number was restored when the egg and sperm fused. According to Weismann's germplasm theory, the hereditary particle called ids (later called genes) situated on idants (now called chromosomes) constituted the germplasm.

Germpore When the pollen grains fall on the stigma, the intine grows out into a tube called pollen tube, through some thin and weak slit or pore called germ pore, present in the exine. In the outer thick exine of the pollen grain, there are a few thin areas which are known as germ pores, germ furrows.

Ghost light *See* bioluminescence.

Gibberellins These are first suspected in a fungus *Gibbererelio fujjikuroi,* an ascomycetes attacking rice. Later gibberelic acid was isolated in 1955. The most important use of gibberelic acid is their spray on grapes to retain them on the vine for a larger period and in much larger size and numbers of fruits. All the plants do not respond equally to gebberillins.

Gigas A mutant having a big size.

Gills In *Agaraicus* (fungus) from the under surface of the pileus suspend a very large number of thin vertical plate-like structure extending from the stipe to the margin of the pileus, these are known as gills or lanuella.

Ginger *Zingiber officinale.* A perennial herb. Extensively used in curries, pickles and various fruit and vegetables' preserves.

Gingelly This oil is obtained from the seeds of *Sesamum indicum*. The seeds yield 45–50 per cent of oil.

Girdle In *Cycas,* the two halves of the leaf trace run around the vascular cylinder in opposite direction, approach each on the side opposite to their place of origin and jointly enter the leaf base. Such leaf bases are termed as girdling leaf traces.

Glabrous When the surface of the leaf is smooth and free from hairs or outgrowth of any kind.

***Gladiolus* sp.** It has underground stem called corm.

Gland Special structures containing some secretory or excretory products from the plant body.

Glandular tissue These are small group of cells with or without a central cavity.

Glaucous When the surface of leaf is green and shining.

Globba bulbiflora American aloe. In this plant, some of the lower flowers of the inflorescence become modified into small multicellular reproductive bodies called bulbils.

Globoid This is found in the aleurone grain, is a double phosphate and calcium and magnesium.

Glumes These are special bracts, small and dry, found in the spikelets of grass family.

Glycine max Fam. Papilionaceae. Soya bean. Important source of vegetable protein and oil.

Glycocaly The outer component of a cell surface, outside the plasmalemma, usually containing strongly acidic sugars hence carrying a negative electric charge.

Glycogen Embedded in the cytoplasm, there are granules of glycogen, protein and volutin and also several oil globules. The protoplast contains several types of granules, the most common of which are of volutin (metachromatic bodies) and of glycogen.

Glycolysis The phase of respiration takes place in the same general way in the presence or absence of oxygen in all types of living cells. The site for this phase of respiration is cytoplasm in the plants. In this phase, glucose molecule splits into two pyruvic acid molecules as a result of a series of reactions.

Glycophyte A plant growing in such a subtratum wherein sodium chloride concentration is more than 0.5 per cent.

Glycoprotein A class of conjugated proteins containing oligosaccharides and protein units.

Glycosmis arborea Fam. Rutaceae. Wild Ilime.

***Gnaphalium* sp.** Stem is provided with dense coating of haris.

Gold mohar *See* Delonix regia.

Golgi bodies In the cytoplasm, there are variety of structures concerned with the metabolic and synthetic activities of the cell. Golgi bodies and centrosomes are found in animal cells. These appear as a collection of tiny plate-like membraneous structures arranged in series and looking like a stack of coins in a plant cell.

Gonidia In *Mucor,* asexual reproduction takes place by spores known as gonidia.

Good speed Clausen and Good speed obtained a periclinal chimaera producing pink flowers. The pink flowered parts have a core of the carmine variety but had its outer portion modified.

Gooseberry Fam. Solanaceae. *Physalis peruviana.*

***Gossypium* sp.** *See* cotton.

Gourds Fruits of Fam. Cucurbitaceae. Snake gourd, bitter gourd, ridge gourd, butter gourd, etc.

Graft An artificially induced vegetative fusion or union of parts from different individuals, the rooted part is called the stock and the part fused to it is scion.

Graft chimeras In a graft, from parts of the surface where the two tissue unite, shoots occasionally arise in the formation of which both species have taken part. The chimeras formed are periclinal or rearely mericlinal. These are called synthetic chimeras or graft chimeras.

Graft hybrids Same as graft chimeras.

Gramineae Family of Monocotyledons. This includes important cereals like wheat, barely, *Pennisetum*, *Sorghum* and others. This family provides us with our food grains of staple food. The grass family. In monocots, the family is the biggest and is distributed throughout the world. This includes about 500 genera and 4000 species. It also constitutes the important plants that constitute the staple foods for human race. The general floral formula is:

$$+ ♂ P_2 \text{ (Iodicules) } A_3 G$$

Grand period of growth Any organ of plant body, in fact every cell first shows slow growth, then it accelerates until a maximum is attained, then it falls off rather quickly and gradually slows down until it comes to a standstill. This whole period is the grand period of growth of the plant or a cell. The time inter from the formative phase to the maturation phase is referred to as grand period of growth.

Grape fruit Fam. Rutaceae. *Citrus paradisi*.

Grapesugar (glucose) Simplest of all the carbohydrates and is found in cane sugar, beet sugar and is formed in the leaf by chloroplasts in the presence of sunlight.

Grasses Common name for Gramineae members like cereals and millets. Most of the fodder which are also important for domestic animals, bamboo, thatch grass and reed as building materials and of sugarcane as a major source of sugar and jaggery is well known. These are important as a raw material for paper industry.

Green-house An enclosure with desired controlled conditions in which young or off season plants are grown.

Groundnut *See Arachis hypogea*. It yields oil from its kernels of which the oil content is about 43–46 per cent an even more. The oil cake is important as animal feed and also fertilizer.

Ground meristem Same as fundamental meristem. The three regions *viz.* promeristem (primordial meristem), the protoderm and the procambial strands from the apical meristem. The rest of the general mass is known as ground or fundamental meristem.

Ground tissue This is the continuous mass of thin-walled parenchyma extending from the sclerenchyma to the pith cavity. In this tissue lie embedded the vascular bundles.

Growth It is a permanent and irreversible increase in size and form attended by an increase in weight, being very slow in plants. In plants, it is defined as an irreversible increase in size commonly accompanied by an increase in dry weight and in the amount of protoplasm. *See* development.

Guanine ($C_5H_5ON_5$) A purine base occuring in nuclei acid. One of the two bases that constitute the DNA.

Guard cells These cells are a part of stomata. They are living and always contain chloroplasts, and their inner walls are thicker and outer walls are thinner. They regular the opening and closing of stoma like lips. A stoma consists of a minute pore–the stomatal opening surrounded by two specialised crecent-shaped or semilunar cells called guard cells.

Gum tree *Acacia senegal* yields gum arabic of best quality for commerce. Gums also occur in mixture of resins.

Gustafsson Isolated several desirable mutants in barley by using X-rays as mutagenic agent.

Gymnosperms These are naked seeded plants *i.e.* in which seeds are not borne enclosed in fruit. They form an intermediate group between cryptogams and angiosperms. Members belonging to this group fall under Spermatophyta (see plants).

Gynadromorphs These organisms are typically female in certain portions of the body and typically male in others.

Gynandropsis pentaphylla A common weed.

Gynandrous When the stamens adhere to the carpels, either throughout their whole length or by their anthers only.

Gynobasic style When the style arises from the depressed centre of the 4-lobed ovary as if from its base or the thalamus.

Gynoecium The pistil is the fourth or the female whorl and its component parts are called carpels. The gynoecium consists of three parts–ovary, style and stigma.

Gynophore The internode between the androecium and the gynoecium is elongated with is the gynophore.

Habitat Ecology deals with the relations between plants and plant-community as they exist in their habitats or places of their living.

Haematoxylon A plant, logwood (an American plant), its heart wood yields the dye haematoxylin.

Haemoglobin An iron protein pigment of blood functioning in oxygen-carbon dioxide of living cells.

Hairs The root bears unicellular hairs while the stem or the shoot bears mostly multicelluar hairs. A dense coating of hairs or presence of still hairs on the body of the plant is always repulsive to animals. Many plants have glandular hairs which secrete a sticky substance. Some plants have stinging hairs.

Halophytes These are plants that can grow in saline soils or water with a preponderance of salts.

Haplo II, III and IV Bridges came across some Drosophila flies with only one dot-like fourth chromosome. Haplo-IV flies have high mortality and reduced fertility. XO flies have also been found but they are sterile. Haplo II and haplo III have not been found in Drosophila.

Haplobiontic In *Ulothrix,* the sporophyte generation is short-lived and is represented by the zygote. Formation of motile or nonmotile spores by reduction division demarcates the end of the sporophyte generation and the beginning of the gametophyte generation. The plant itself is a gametophyte and the sporophytic generation is inconspicuous and short-lived. Hence the alternation of generation in Ulothrix is haplobiontic.

Haploid (haploose-single) The gametic number of chromosomes (n); in an organism whose somatic nuclei have a single set of chromosomes.

Haploid number The lowest multiple number in a polyploid series. This is also the basic number.

Haplomitosis A type of primitive mitosis in which the nuclear granules form into thread like masses rather than clearly differentiated chromosomes.

Haplont A plant with haploid somatic cells.

Haplosis Reduction of the chromosome numbers to half during meiosis.

Hapteron In *Spyrogyra (Algae)*, usually all the cells are smaller in shape, size and function but in some species the basal cell called hapteron is colourless and bears numerous rhizoidal processes as in *S. jogensis.*

Haptotropism It is the movement of an organ induced by contact with a foreign body. Twining stems and tendrils move like this. This is a kind of movement stimulated by contact with a foreign body. Also called thigmotropism.

Hard bast The phloem fibres in the secondary phloem are the hard bast.

Hard wood A broad leave tree that provides heavy and relatively hard-wearing close grained timber of the wood got from this tree. Both tropical (*e.g.* teak) and temperature (beech) species of hardwood exist.

Harland Hardland and Beasly independently synthesised amphidiploid by colchicines treatment of the sterile hybrid *G. arboreum* X *G. thuberi.*

Hastate When the two lobes of a sagittate leaf are directed onwards.

Haustorium To absorb food from the host plant, a parasite produces certain special roots or haustoria. These are also called sucking roots.

Head of flower *See* capitullum.

Heart wood The parenchyma cells associated with xylem are killed but before dying they produce certain antiseptic substances like tannins, gums and inorganic crystals. Some ogany and ebony certain dyes are also added. Also *See* duramen.

Hedera helix *See* ivy, climbers.

Helianthus annuus Fam. Compositeae. Sunflower. A very important oilseed crop.

Helianthus tuberosus Fam. Compositeae. Jerusalem artichoke produces tuber like potato.

Helicoid This is a one-sided cyme in which successive branches develop on same side, forming a sort of helix.

Heliotaxis Orientation movement of an organism in response to the stimulus of sunlight.

Heliotropism Also called phototropism. It is a kind of movement of plant organs in respect to incidence of rays of light.

Hematochrome The red colour in the eye-spot or stima in *Chlamydomonas pluristigma* (Algae). The red colour is due to this pigment.

***Hemerocallis* sp.** Fam. Liliaceae. Day-lily.

Hemiascomycetes *See* Plant Pathology.

Hemicryophyte A plant having buds at the soil surface and protected by scales, snow or litter.

Hemiphragma heterophyllum This is heterophyllus plant showing adaption to two different conditions of environment, aquatic and aerial.

Hemizygous A diploid organism having only one copy of a gene in place of the normal two.

Hemp *Cannabis* sp., is the source of a narcotic used in the form of smoke. The Indian hemp is *Crotolaria juncea* and is vastly used as a green crop for green manuring of cultivated fields. The Deccan hemp *(Hibiscus cannabinus)* and bowstring hemp *(Sansevieria roxburgiania)* also yield fibres.

Hepaticae Bryophyta is usually divided into two classes namely Hepaticae or liverworts and Musci or mosses.

Herbs These are small plants with a soft stem. They may vary from a few millimetres to a metre or so in height.

Heredity The transmission of parental qualities, expressed or latent, to the progeny.

Heritability The portion of observed variability which is due to inheritability, the remainder being due to environmental cause.

Heritable variations These may be caused by either recombination of genes or by mutations.

***Heritiera* sp.** A mangrove plant.

Herkogamy When some kind of barrier stands between the stamens and the pistil of the same flower. A hood covering the stigma is a common form of such barrier.

Hermaphrodite Same as bisexual.

Hertwig Oscar Hertwig, a German zoologist, showed that in the sea urchin, fertilization involved a union of the nuclei of one egg and one sperm.

Hesperidium This is a superior, many celled fruit (fleshy) with axile placentation. In this the endocarp projects inwards forming distinct chambers, and the epicarp and the mesocarp, fused together form the separable skin of the fruit.

Heterobeltiosis To describe the improvement of heterozygote over the better parent of the cross.

Heterochromosomes Chromosomes distinguished by special peculiarities of behaviour, form or size.

Heteroecious An organism requiring more than one host for completing its life cycle. Such fungi (rusts) which require two distinct host plants to complete their life cycle while others are autoecious.

Heterogametes In Cryptogams, sexual reproduction takes place by the fusion of two similar gametes, called isogametes. But sometimes two dissimilar gametes, called heterogametes, differentiated into make and female fuse together for the sexual reproduction.

Heterogametic sex When the male gametes are produced in two kinds.

Heterogamy Union of two unlike gametes.

Heteromixis Sexual reproduction which involves the fusion of genetically different nuclei.

Heterophyllus Having more than one form of foliage leaves on the same plant.

Heterophylly When plants bear different kinds of leaves on the same plant.

Heterophytes Non green plants cannot prepare their own food and thus they draw it from different sources. Such plants are called heterophytes or heterotrophic plants. They are either parasites or saprophytes.

Heterosis The increased vigour often exhibited by hybrid individuals.

Heterosis theory One explanation is based on the assumption that hybrid vigour is entirely due to the bringing together in the first generation hybrid of a large number of favourable dominant genes contributing to vigour. Second explanation is that hybrid vigour result from heterozygosity. According to this hypothesis called the 'heterozygosity hypothesis' proposed independently by Shull and by East in 1908, heterozygous alleles exert a complementary physiological action leading to increased vigour over the homozygous. Consequently the most vigorous individual is the one with the greatest number of heterozygous alleles.

Heterospores Production of more than one type of spores by a single plant *i.e.* microspores and megaspores.

Heterosporous In Silaginella, the microsporophyll bears in its axil a microsporangium with usually 16 microspore mother cells. All of them undergo reduction during and give rise to 64 small microspores with groups of four. Silegialla is thus heterosporous. The main plant body in ferns' *Pteridophyta* is the sporophyte and reproduces asexually with the formation of spores. The spores in ferns are of one type and this condition is known as homosporous, in contrast to some other Pteredophytes where there are two types of spores–small and large, and are termed as heterosporous.

Heterostyly Some plants bear flowers with two different forms–one form is long stamens and a short style and other form bears short stamens and a long style. This is called dimorphic heterostyly. Similary, there may be cases of trimorphic heterostyly.

Heterothallic In Alga, conjugation normally takes place between the gametes from two different filaments only, hence *Ulothrix* is physiologically heterothallic. Also *See* Plant Pathology.

Heterotrophs Such organisms are unable to trap any primary source of energy and have to depend upon the respiratory energy (the secondary energy) for the synthesis of organic compounds. The heterotrophs are either parasites when they obtain their carbon from living organism or saprophytes when they get their carbon from the dead remains of the dead products of living organisms.

Heterozygote (hetero-different; zygote-egg) An organism derived from the union of gametes of disimilar genetic constitution.

Heterozygous Having unlike alleles at corresponding loci of homozygous chromosomes. An organism may be heterozygous for one of several genes.

Heterozygous translocations These can be readily detected by the cross shaped pairing configurations. Such cross shaped configurations permit a termination of the points at which these chromosomes had broken to produce the translocations.

Hevea brasiensis Rubber is obtained from the latex of this big tree. Latex is collected by tapping the bark.

HFr High frequency recombination strain of *Escherichia coli* resulting from integration of F plasmid in the bacterial chromosomes.

Hibiscus subdariffa Fam. Malvaceae. Roselle plant. Useful as a fibre; and fruits can be used for making fruit jelly. *H. canabinus* is the Deccan hemp.

Hill reaction In the Photosynthesis Hill (1935) found that isolated chloroplasts suspensions when illuminated posses the power to reduce ferricions and at the same time produced oxygen. The reaction can be shown as:

$$4\,Fe^{++} + 2H_2O \xrightarrow[\text{Chloroplast}]{\text{Light}} 4\,Fe^{++} = 4\,(H)^{+} + O_2$$

Hilum On one side of the seed, lying above its projected end, a small oval depression in the hilum. The place where funicle is attached to the nucellus is known as hilum.

Histology A detailed study of cellular structure of various plants' parts with the help of microscope is histology.

Histone Any of several proteins that can combine with DNA to form chromosome.

History of biology Although the world biology (Greek Bios-life: logos-discourse, science or study) was coined independently by two famous naturalists–Lamark and Triveranus only in year 1802, the study of plants and animals is as old as man himself. It had its roots in antiquity.

Holdfast In Ulothrix, it fixes itself to any hard object in water, by the basal elongated colourless cells called the holdfast. Commonly the plant remains to the sub-stratum by the lowermost colourless elongated cell.

Hollyhock Fam. Malvaceae. *Althea* sp. Ornamental plant.

***Holmskioldia* sp.** Lady's umbrella. A medicinal plant.

Holorrhena antidysentarica Fam. Apocynaceae. Medicinal plant.

Homeomorph An organism which exhibits a superficial resemblance to another organism even though they have different ancestors.

Homogametic sex The sex which produces only one type of gametes with regard to the sex chromosomes. When the like kind of gametes are formed as in females. But there are instances where there is the heterogametic sex and the male is the homologous etc.

Homologous chromosomes Chromosomes in which the same gene loci occur in the same sequence. These occur in pairs, one derived from each of the two parents normally, (except in case of chromosomes associated with sex such as X and Y chromosomes), morphologically alike and bearing the same gene loci, each member of such a pair is the homologous of the other.

Homologous pair In the somatic cells (or the vegetative cells) of most organisms that reproduce sexually, chromosomes are found in pairs, of which one membranes comes from the mother and other from the father.

Homology Study of morphologically modified organs of plants.

Homologues Shortly after the chromosomes appear as long thin threads in the leptonema stage of meiosis, the two members of each pair chromosomes (*i.e.* homologous chromosomes or homologoues), which are similar in shape and size and which contain similar (or alleles) genes, approach each other.

Homospores In the asexual reproduction of *Pteridophyta Adiantump eridium*, the spores are of one type and the constitution is kown as homosporous, in contrast to some other Pteridophytes where are two types of spores–small and large and are termed as heterosporus.

Homozygote An organism derived from the union of similar genetic constitution.

Homozygous Having like genes at corresponding loci on homologous chromosomes. An organism can be homozygous for one, several or all genes.

Homozygous translocation Such translocations may not exhibit any cytological peculiarities. They behave like the standard (or normal) types and often have regular pairing at meiosis.

Hooke Cells were first seen by Robert Hooke under his primitive microscope in 1665.

Hordeum vulgare Fam. Gramineae. Barley, is a staple food also used as feed for animals, the straw is also used for animal fodder.

Hormogonia The *scillotria* (Algae) reproduces only vegetative there being no sexual or asexual reproduction. During reproduction, the trichome breaks into a number of many celled pieces called hormogonia.

Hormone Certain organic products formed in very minute quantities as a result of metabolism inside the plant body have a profound influence on the growth of the plant organs and on various kinds of tropic movements exhibited by such organs, they also have a marked effect on certain physiological processes in the plant. These are known as hormones or growth regulating substances.

Hornwort *Ceratophyllum* sp. A submerged plant.

Horowitz The work of Horowitz, Benzer, Beadle, Tatum, Lederberg, Zinder and others on the bread mould *Neurospora*, the colon bacterium *Escherichia coli*, the typhoid bacterium Salmonella and bacteriophages has been responsible for the rapid progress made in recent years in the fields of physiological and biochemical genetics.

Host pathogen relationship A disease is the result of an interaction of genes governing resistance in the host with those governing pathogenicity in the parasite.

Hot water method of emasculation Mass emasculation can be done by the hot water method in rice and sorghum.

***Hoya* sp.** Commonly known as wax plant. Medicinal value.

Hugo de Vries Hugo de Vries in Holland, Carl Correns in Germany and Erich Tscermak in Austria independently began to hybridize plants and unknown to another, obtained results similar to those obtained by Mendel though none of them had seen Mendel's paper before beginning his experiment.

Hull The superiority of the heterozygote ($A^1 A^2$) over the homozygotes ($A^1 A^1$ or $A^2 A^2$ was termed over dominance by Hull (1945).

Hutchinson Hutchinson and Stephens suggested that the cultivated Asiatic cottons were carried by an early civilization to the new world and that natural hybridization of the cultivated crop with a neighbouring American species gave rise to the first amphidiploid.

Hyaloplasm The fluid portion of endoplasm.

Hybrid The first generation offspring of a cross between two individuals differing in one or more genes; the progeny of a cross between species of the same genus or of different genera.

Hybridization The crossing of individuals of unlike genetic constitution.

Hybridization in self-pollinated crops Rice, wheat, barley, groundnut, pea, tobacco, tomato and brinjal are examples of highly self-pollinated crops in which natural crossing is hardly one per cent. Sorghum and cotton are the crops of cross pollinated nature in which natural crossing under most environmental conditions is not less than 5 per cent and very often may be more than 10 per cent.

Hybridization technique It consists of removing the anthers before the shedding of pollens, collection of viable pollen from the male parent, and transferring it to the receptive stigma of the emasculated plants. Removal of anthers is done one day before the pollen is ripe.

Hybrid vigour The increased vigour, growth, size, yield of function of a hybrid progeny in relation to the average of the parents.

Hydathodes These are found mostly in the aquatic plants or are those herbaceous plants in moist place. These are also known as water stomata in water, water pores are present at the margin of the leaf opposite the vein endings.

Hydrophytes Plants mainly growing in water as well as those inhabit swamps or marshy habitats.

***Hydrilla* sp.** A plant growing in submerged conditions.

Hydrophily the plant is dioecious and submerged. The male flowers get detached from the small spadix of the male plant and float on water. Each female borne on a long stalk by the female is brought to the level of water. The anthers burst and the pollen grains are distributed on the female flower.

Hydrophytes This is an ecological group of plants. These are the plants that grow in water or in very wet places. These plants show adaptations to their habitats, such a large leaf, projecting stomata, and thin cuticles for efficient transpiration and short root system.

Hydroponics It is the technique of growing plants directly in normal culture solution including the essential trace elements without the use of soil.

Hydrosphere *See* Soil Science.

Hydrostatic In plant cell, osmosis continues until the hydrostatic pressure due to the accumulated flow of the solvent has attained a value sufficient to stop further flow.

Hydrotropism The movement of an organ in response to the stimulus of moisture.

Hygrometer An instrument that measures the relative humidity of the atmosphere. The most common type uses a human hair which expands or contracts as the relative humidity increases or decreases while another type uses a lithium chloride wire, the resistance of which varies with the changing humidity. A continuous record, a hygrogram showing changes of R.H. of the atmosphere over a given period of time.

Hygrophyte A plant adapted to a plentiful and regular moisture supply but not to fully water logged condition. Plants of tropical rain forests and humidity loving ferns and mosses are the examples of such plants.

Hygroscopic water *See* Soil Science.

Hyloplasm The ground substance of cytoplasm which is colloidal and contains enzymes for glycosis and structural materials such as sugar, amino acids, water, vitamins, nucleiotides etc., commonly known as cytoplasmic matrix, nutrient soup or cell sap.

Hymenium In fungi, the haploid cells of the trama curve outwards on either side of the gill and terminate in a layer of small rounded cells–the subhymenium and layer of club-shaped cells–the hymenium.

Hypanthodium When the fleshy receptacle forms a hollow cavity with an apical opening guarded by scales, and the flowers are borne on the inner wall of the cavity, the inflorescence is a hypanthodium.

Hyperparasitism Many parasites may be infected with smaller parasites of their own and in turn may support still smaller ones. The later form of parasitism is known as hyper parasitism.

Hyperplastic disease This type of disease is characterised by the formation of tubercles, tumours or galls, as in crown-gall of fruit trees due to phytomonus and tumefacience.

Hypertrophy Abnormal enlargements of varied characters are found in plants attacked by the fungi, resulting in the formations of tumours, knots, galls or warts. In growth of a plant, there is a great coordination between the various physiological activities, which results in the orderly development of tissues and organs. Growth without coordination results in hypertrophy, the formation of abnormal growth such as galls and tumours.

***Hyphacene* sp.** Fam. Palmae. This plant shows dichotomous branching which is rare in this family.

Hypnospore In Algae, at the advent of normal conditions, the daughter protoplasts develop flagella, contractile vacuoles and an eyespot, and start leading normal life as soon as they escape out of the jelly. Rarely the cells of the palmella stage develop into thick walled hypnospores.

Hypobasal In Pteredophyte, the wall between the archegonium and prothallus is the basal wall, which divides the oospore into halves–An anterior or epibasal lying towards the notch of the prothallus or the hypobasal lying towards the posterior end.

Hypocotyl The portion of the axis lying immediately below the cotyledons. Germination of this kind is said to be epigeal.

Hypodermis A few external layers of collenchyma and sometimes sclerenchyma.

Hypogeal germination The cotyledons remain in the soil or just on its surface.

Hypogyny in a typical flower, ovary occupies the highest position on the thalamus, while the stamens, petals and sepals are separately inserted successively below the ovary.

Hyponasty The condition, when leaves at first remain closely coiled because of more rapid growth on the undersurface.

Hypophysis In the embryo of angiosperms, the upper-most or the basal cell of the suspensor which is attached at the mycropylar end of the sac enlarges very much and draws nourishment for the suspensor. The lowermost cell of the filament which lies next to the embryo is known as hypophysis.

Hypophysis cell The basal cell of the suspensor often enlarges and acts as an absorbing organ, while its terminal cell is called hypophysis, cell divides and gives rise to the apex of the radicle.

Hypostatic *See* epistatics.

Hypsometer A scientific instrument that is used to calculate altitude or atmosphere pressure by measuring the temperature at which water boils. As the boiling points varies with the pressure (this can be calculated from the temperature at which boiling occurs. An equation is then used to calculate the altitude. The accuracy of this method is not very high.

Ibris **sp.** Fam. Cruciferae. Candytuff. Ornamental.

Idiogram A diagramatic representation of the complete chromosome set of an individual, arranged according to (decreasing) size and/or accepted numbering system. Idiograms may be drawn from suitable microscopic preparations, or these may be photographed and various chromosomes be arranged in order (and in homologous pairs in diploid organisms).

Idiosome A hypothetical unit of cell such as the region of modified cytoplasm surrounding the centrosome.

Ids *See* germplasm theory.

Illegitimate crossing over Same as reciprocal translocation.

Imbibition Sachs (1974) suggested that water moves along the walls of xylem vessels (and not through their cavities) as a result of imbibition of water by the solid particles of vessel walls.

Imbricate When one of the sepals or petals in internal by being overlapped on both the margins, and one of them is external and each of the remaining ones is overlapped on one margin and it overlaps the next on the margin and it overlaps the next one on the other margin. *See* aestivation.

Immunity Complete resistance to any disease or insect-pest.

Imparipinnate When the midrib of the pinnately compound leaf directly bears the leaflets, it is said to be unipinnate. In it the leaflets may be even in number (paripinnate) or odd in number (imparipinnate).

Impermeable membranes Such fine membranes are found in animal and plant bodies which can allow certain fluids or distinct density to pass through them and not other and all.

Impervious *See* Soil Science.

Impicac Fam. Rubiceae. *Psychotria sp.*, yields emitine.

Importance of linkage In the scheme of improvement of plants, the breeder aims at recombination of desirable characters through hybridization, but in the achievement of this aim, difficulties often arise on account of linkage.

Importance of mitosis Heredity is maintained from cell to cell. The longitudinal division of the chromosomes assure that the daughter cells have quantitatively and qualitatively the same genetic constitution as that of the original cell from which they arose. Mitosis is thus the mechanism which maintains heredity, whether it be from cell to cell or organism to organism.

Inbreds By self pollination followed by selection, homozygous plants can be obtained and inbred lines with desirable characters 'fixed' in homozygous condition can be developed for these.

Inbred line A line showing a considerable degree of homozygosity as a result of continued inbreeding and selection.

Inbreeding Breeding closely related organisms; in plants usually be self pollination.

Incipient plasmolysis The stage of plasmolysis at which the first sign of shrinkage of protoplasm from the cell wall becomes visible is called incipient plasmolysis.

Incompatibility The inability of pollen to bring about fertilization because the pollen tubes are arrested in the style and prevented from reaching the ovules.

Incomplete dominance The production of an effect by two different alleles that is intermediate to the effect produced by the same allele in a homozygous condition.

Incomplete linkage Linkage of the segregating genes of a dihybrid leads to deviations from the test ratio 1 : 1 : 1 : 1. The magnitude of such deviation depends upon the frequency with which recombinations occur (*i.e.* depends on the strength of linkage). In such linkage, the F_2 generation is not represented normally by the expected phenotypes, but along with the two major types (linked) some of the in between do occur, through in a very small number.

Independent character Characters governed by genes showing independent assortment.

Independent gene Genes which show independent assortment.

Indigo Fam. Leguminoseae. *Indigofera indica.* An industrially important plant.

Induced mutation Induce mutations have been put forth only after the works of Muller, H.J. after 1927.

Induced polyploids When it was observed that polyploids, in general were larger and more vigorous than the diploids, interest in the induction of polyploids increased. To meet this end, various kinds of agents like temperature, chemicals, X-rays and gamma-rays etc. were used to evolve induced polyploids. Of these colchicine was very frequently used.

Inducer A substance inactivating a represser.

Inducium A fern, each sorus is a group of sporangia covered over by a kidney s-shaped shield called the inducium.

Indusium In pteredophyta, it is the covering sheathe of a sorus.

Inert alleles The active genes are designated with capital letters of whose recessive contributors designated by small letters and do not in any case influence the concerned character in plant *e.g.* height.

Infection Lysis has three distinct phases, first phase being infection, second synthesis and the third lysis proper.

Inflorescence The branch system of the flora region bearing a group of flowers. It is the floral bearing part of a plant.

Inhibitory gene Such a gene has no phenotypic expression of its own and its presence can be judged only by its effect on a certain particular gene. In the F_2 of a dihybrid cross this gives a phenotypic ratio of 13 : 3.

Innate The anther is said to be innate or basifixed when the filament is attached to the base of the anther.

Integument The two layers of which the seed coat is made up of are known as integuments. The nucellus is covered by one or two cellular covering sheaths known as integument.

Interaction of two paris of alleles Such interaction of independently transmitted pairs of alleles that are not independent in their expression. A classical case of two pairs of alleles affecting the same characteristic and producing the in the F_2 four different phenotypes in the ratio of 9 : 3 : 3 : 1.

Interchange An exchange of non-homologous terminal segments of chromosomes.

Interfascicular cambium At first a portion of each medullary ray in a line with the cambium becomes meristematic and forms a strip of cambium called the interfascicular cambium. It is a cambium when the left over of the apical meristem is primary in origin is also known as intrafascicular cambium.

Interference The reduction in likelihood of a second crossover closely adjacent to another. Interference ordinarily increases with decreased distance.

Intergenic suppression The restoration of a suppressed function or character as consequence of a second mutation located within the same gene as the original mutation.

Interkinesis or Interphase. Resting stage between the first and second divisions of meiosis, or the resting stage between two mitoses.

Interpetiolar stipules These are the two stipules that lie between the peticles of opposite or whorled leaves, thus alternating with the latter.

Interphase The stage of cell life during which the cell is not dividing. It is sometimes known as resting phase though at this stage important changes occur in the nucleus and it directs all its activities of life.

Intersexes Intersexes are sterile individuals intermediate females and males but are different from gynandromorphs which are typically female in certain proportions of the body and typically male in others.

Interstitial chiasmata If the chiasmata are found anywhere along the arms of chromosomes.

Intervening sequences A region of a gene that is transcribed but is not included in the final transcrip of a ribonucleic acid.

Intine The inner wall of a pollen grain. Also *See* exine.

Intramolecular respiration Very little energy is released by the process of anaerobic respiration which is represented by: $C_6H_{12}O_6 \rightarrow 2C_2H_5OH + 2CO_2$ + 28 cal. (sugar ethyl alcohol + carbon dioxide + 28 cal. of energy). It is otherwise known as intramolecular respiration.

Introgressive hybridization The spreading of genes of a species into the gene complex of another due to extensive back crossing between numerically dissimilar population preventing formation of a single stable population.

Intussusception growth Growth of the cell wall in surface area *i.e.* its increase in size, takes place in the early stage of the cell and is due to stretching of the cell-wall in one or more directions accompanied by an addition, within the original wall, of new solid particles secreted by the protoplasm. This method is called growth by intussusception. When the cell grows, the cell wall is stretched and new cellulose particles are secreted by the protoplasm intercalated or fitted in the interparticulate spaces of the existing wall. This increases the surface area of the wall and this type of growth is found on early stages of the growth of the cell.

Inulin It is a soluble carbohydrate and occurs in solution in the cell sap, when required for nutrition it is converted into a form of sugar (fruit sugar or fructose).

Inversion The reversal of one segment of a chromosome relative to the rest of the chromosomes.

Inversion heterozygotes In individuals in which there is an inversion in one of the two homologous chromosomes, true synapsis is not possible by simple linear pairing. Inversion heterozygotes synapse by forming loops.

Involucular leaves In the Bryophyta, the apices of the leafy shoots surrounded by the sex organs are borne at the apices of the leafy shoots surrounded by the special leaves called involucular leaves.

Involucre In an umbel, there is always a whorl of bracts forming an involucre.

Ion The mutation inducing radiations are of two kinds–(1) Ionizing radiation, and (2) Non-ionizing radiations.

***Ipomea* spp.** Fam. Convolvulaceae. *Ipomia batas* (sweet potato), *I. reptens* (water bine weed), *I. paniculata* (medicinal), *I. purpurea* (morning glory), *I. palmata* (railway creeper), *I. grandifloral* (moon flower) are used in various ways such as vegetables, medicines and ornamental.

***Iris* spp.** This genus has petaloid style.

Irritability The capacity of plants and their particular organs to receive stimuli from outside and to respond to them. Living organisms are sensitive and show response to stimuli such as heat, light, touch, chemicals etc. It is a physical nature of protoplasm that it shows such sensibility.

Isoallele An allele that carries mutational alterations at the same site.

Isogamete Gametes not differentiated as to sex. Fusing isogametes are similar but, in some species are chemically differentiated.

Isogamy When two fusing gametes are similar in shape, size and behaviour. Such gametes are formed by the fission of the protoplast in the same manner as in asexual reproduction.

Isobilateral leaf Such leaves have upper and lower epidermis having stomata on both sides.

Isotopes Isotopes are chemically identical substances having the same number of protons but different numbers of neutrons.

Ivy These plants have climbing roots. To ensure a foothold, such roots secrete a kind of sticky juice which quickly dries up in the air.

Ixora coccinea Fam. Rubiaceae. An ornamental plant.

Jaculator A curved hook present in twining stem.

Jalap Fam. Convolvulaceae. *Operculina terpethum.* Medicinal plant.

Janssens He discovered that, during meiosis, corresponding chromosomes from mother and father are held together at certain points called 'chiasmata' and put forward the theory that exchange of the segment between chromatids of material and paternal chromosomes took place at these points.

Jasmine Fam. Nictyginaceae. *Nyctanthes* sp. and *Cestrum* sp. (queen of the night) and *Jasminum* sp. These are all sweet scented plants.

***Jatropha* sp.** Bears fruits called regma, which form fruits from a syncarpous pistil, 3 to many chambered.

Jerusulum thorn Fam. Mimoseae. *Parkinsonia aculeata.* A weed.

Jew's slipper Fam. Euphorbiaceae. *Pedilanthus* sp. Bears cyathium type of inflorescence.

Johannsen A Danish botanist Whelhelm Johannsen introduced the word 'gene' for the material present in the germ cells which determine the many properties of an organism. He also introduced the work 'genotype' for the heredity of an individual plant or animal. He also developed the concept of 'pure line'.

Jones Jones (1917) put forward the hypothesis that each chromosome that contain some favourable recessive genes and some bad dominant genes. Several crossovers would be necessary to obtain all of the dominant alleles in one gametes.

J-shaped chromosome The shape of a chromosome is largely determined by its centromere if the centromere is located near the end of a chromosome, it will be rod shaped and if it is located at a small distance inside the chromosome, the chromosome is J-shaped.

***Juncus* sp.** Rush. Bears monochasial cyme or uniparous inflorescence.

Jussiaea repends Is an aquatic plant, having floating branches which develop special types of roots known as respiratory roots.

Jute Fam. Tiliaceae. *Chorchorus capsularis* and *C. olitorius*. Its fibres are extensively used for making gunny bags, cheap rugs, carpets, cordage, hesian (course cloth).

Kalanchoe verticilata A series of adventitious (foliar) buds are produced on the leaf margin, each at the end of a vein, these buds are produced on the leaf margin, each at the end of vein, these buds grow up into new plants.

Kappa particles *See* Paramecium.

Karpechenko An intergeneric amphidiploid was produced by Karpechenko by crossing the radish *(Raphanus sativus,* 2n-18) with a cabbage *(Brassica oleracea,* 2n-18).

Karyokinesis Mitosis.

Karyology The biology of the nucleus.

Karyolymph Nuclear sap.

Karyotype Character of a chromosome complement with regard to the number and shape of the chromosomes.

Keeble Keeble and Pellew proposed the dominance hypothesis.

Keel In the papilionaceous flower having five petals on which the outermost one is the largest and known as the standard or vexillum, the two lateral ones are known as the wings or aloe and the two innermost ones are the smallest and are together known as the keel or carina.

Ketoglutaric acid After the formation of citric acid (Krebs cycle) reactions follow one after another in quick succession, and CO_2 is released from some of these reactions. Soon this acid appears in the cycle.

Kihara Kihara in Japan succeeded in sympathising a hexaploid wheat by crossing the emmer wheat *(T. dicoccoides)* with *Aegilops squarrosa* (2n-14) and then doubling the chromosomes of the hybrid. The synthesised species resembled the common bread wheat and when crossed with it, gave fertile offspring with regular meiosis.

Kinetochore Same as centromere.

Kinetics of growth In the initial period during which internal changes preparatory to growth take place, then a phase of increasing growth rate comes. After this, the phase where the growth rate gradually diminishes and a point comes when the individual reaches maturity and the growth stops.

Knop's normal culture solution Potassium nitrate-1 gm., acid potassium-1 gm., magnesium sulphate-1 gm, calcium nitrate 4 gms, Ferric chloride solution.-a few drops, water-1000 c.c. minute doses of 'trace' elements are to be added.

Kohl-rabi or knol-kohl Fam. Crucifereae. *Brassica caulorapa.* Used as vegetable.

Kreb's cycle During aerobic respiration with the access of O_2, the pyruvic acid is completely oxidised by a group of enzymes (oxidases) into CO_2 and H_2O. Maximum amount of energy is released by this process. A series of reactions take place in a cyclic order, called the citric acid cycle or krebs cycle. First series of reactions bringing about complete oxidation of pyruvic acid are taking place in the outer membrane of mitochondria. These complicated but systematic series of reactions are cyclic in nature like many other biochemical reactions and are known as Kreb's cycle.

Labiateae A family of angiosperms. It abounds in volatile aromatic oils, which are used in perfumery and also as medicines.

Laburnum (Indian) Fam. Caesalpinieae. *Cassia fistula.* An ornamental tree with flowers.

Lagenaria siceraria Fam. Cucurbitaceae. Bottle gourd. An important vegetable plant.

Lagerstroemia **sp.** Crepe tree. An ornamental hedge. Produces winged seeds.

Lamark Jean Baptiste de Lamarck, a French biologist (1744-1829).

Lamark's theory He was of the opinion that environmental changes causes modifications in organisms and that such modifications are transmitted to subsequent generations. He believed that environment acts directly on plants and indirectly on higher animals.

Lamellae *See* gills.

Lanceolate When shape of the leaf is like a lance.

Laportia **sp.** It is a nettle having stinging hairs.

Larkspur Fam. Ranunculaceae. *Delphinium* sp. Ornamental.

Latex The latex vessels are formed by the dissplotion of the walls between rows of latex cells and by the fusion of their branches.

Latex cells Although such branches like the latex vessels are really single land independent units. They branch profusely through the parenchymatous tissue of the plant, but without fusing together to form a network.

Latex vessels These are row of more or less parallel ducts connected with one another by the fusion of their branches, forming network. These vessels are found in poppy (Papavaraceae) family.

Lathyrus odoratus Flower colour is inherited in the F_2 by showing 9 : 7 ratio (complementary genes).

Lathyrus sativus Fam. Papilionaceae. Khesari. It is used as pulse but is mostly a weed.

Laticiferous tissue It consists of much branched, elongated ducts which contain a white milky fluid called latex.

Latin square design An experimental field design having as many replications as treatments so that the layout forms a square.

Lattice design The experimental field design in which the number of varieties to be tested form a square of the same number and numbered serially. These are arranged in the same sequence in equal number of squares.

Lavender Fam. Labiateae. *Lavandula* sp. Yields lavender oil.

Law of independent assortment Mendel's Second Law of inheritance or the Law of independent assortment is that the segregation in one pair of alleles is independent of the segregation in any other pair of alleles.

Law of segregation This is Mendel's first Law of Segregation and is expressed as that allelic genes in a hybrid do not blend or contaminate each other but segregate and pair into different gametes.

Leaf Leaf may be regarded as a flattened, lateral outgrowth of the stem or the branch developing from a node and having a bud in its axil. Mostly the leaves are so borne on the stem that they are placed horizontally or almost so.

Leaf functions (1) Manufactures food (2) Interchange of gases between the atmosphere and plant body (3) Evaporation of excess water through the leaf (4) Fleshy leaves store water and food (5) In few cases leaf produces buds on it for vegetative propagation (6) Leaf gives protection to small bud in its axil and (7) Modified leaves have special functions.

Leek Fam. Liliaceae. *Allium porrum.* Used as vegetable.

Legume This is a dry, one chambered fruit developing from a simple pistil and dehiscing by both the margins. It is also called a pod.

Leguminoseae A family of angiosperms. Has herbs, shrubs and trees in it. Legume family. The floral formula is:

Subfamily Papilionaceae $+ ♂ K(5)\ C_1 + 2x\ A_1 + (9)\ \underline{G1}$

Subfamily Caesalpiniodeae $+ ♂ K_5 Cs A_{3+3+3}\ \underline{G1}$

Subfamily Mamosoideae $\oplus K(5) C_5 (3) A_2\ (2)\ \underline{G1}$

Lemina The flat green floating blade (stem) of duckweed.

Lemma In the small spike (florets) there are two small empty glumes at the base and just above them a flowering glume called lemma.

Lemon Fam. Rutaceae. *Citrus* sp. Many species exist. A fruit very rich in citric acid.

Lemon grass *See Cymbopogon* sp.

Lenticular transpiration In woody plants, some transpiration takes place also through the lenticels and is termed as lenticular transpiration.

Lentil Fam. Papilionaceae. *Lens esculentas.* An important pulse crop.

Leonurus sibiricus Fam. Labiateae. A waste land plant.

Lepidium sativum Fam. Crucifeaeae. Garden cress, ornamental.

Leptonema (or leptotene). One of the stage of meiotic prophase when chromosomes appear as very long and slender threads, but because of their extreme length, are not individually visible.

Lethal genes Genes which in the homozygous state, have such a marked deleterious effect that such homozygous organisms are inviable.

Lettuce Fam. Compositeae. *Lectuca sativa.* A leafy vegetable. The water lettuce is *Pisticia* and grows as submerged condition plant.

Leucas linifolia* and *L. aspera Fam. Labiateae. Both are weeds growing on waste lands and have medicinal value.

Leucoplast It is one of the plastids. According to their colour, the plastids are three types *viz.* leucoplasts, chloroplasts and chromoplasts. One form of plastid can change to another.

Lichens These are associations of algae and fungi, and commonly occur as thin round greenish patches on tree trunks and old walls.

Life cycle The living organisms possess a definite life cycle. After birth, every organism grows on nutrition and metabolic activities, becomes mature, reproduces, enters old age and ultimately dies.

Light and dark reaction The initial phase of photosynthesis, the splitting of water (H_2O) into oxygen and hydrogen under the influence of chlorophyll in the presence of light (source of radiant or solar energy).

Light effect A certain intensity of light maintains the healthy condition of plants; absence of light makes plants soft, weak, brittle, slender, long and drawn out, pale green or pale yellow in colour and sickly in appearance; such plants are called etiolated.

Lignification When lignin is incrusted in the original cellulose wall, the process is known as lignification.

Lignin The secondary thickening of the cell wall and tracheid takes place after they have grown considerably, and attained their full dimensions, their walls begin to thicken. The thickening in them is due to the deposit of hard and chemically complex substance, called lignin.

Light reaction During this phase, light energy absorbed by chlorophyll is temporarily converted into chemical energy or ATP (Andenosin triphosphate) and reduced NADP *i.e.* NADP (H_2). Although light reaction is a complicated system of several reactions occurring simultaneously yet to understand it clearly and easily. It has been divided into three parts–(1) protolysis, (2) photophospolysation, and (3) reduction of NADP.

Ligulate Also called strap-shaped. When the corolla forms into a short narrow tube below, but is flattened above like a strap.

Ligule In Gramineae. A hairy structure called ligule is present at the base of the leaf blade.

Lilac (Persian) The leaves in this plant are oblique.

Liliaceae A family of angiosperms. Have many useful plants in it. Floral formula: ⊕ ♂ $P_{(3+3)}$ $A_{3+3 \text{ or } 3+O}$ $\underline{G(3)}$

Lily Fam. Liliaceae. *Lilium* sp. An ornamental plant. Day lily is *Hemerocallis* sp. Ester lily is *Amaryllis* sp. Pin-cushion lily is *Haemanthus* sp. Zephrilily is *Zephyranthes* sp. and Eucharis lily is *Eucharis* sp.

Limb In the polypetalous corolla, each petal may sometimes be narrowed below, forming a sort of stalk known as claw, and expanded above; this expanded portion is the limb.

Limnophilla heterophylla This is a plant amphibious in nature and habitat.

Limonia acidissima Fam. Rutaceae. Elephant apple.

***Lineria* sp.** Fam. Linaceae. It has a personate or masked corolla.

Linkage Association of characters due to location of genes in proximity on the same chromosome.

Linkage group The number of linkage groups will equal to the haploid number or chromosomes which the species possesses. Thus, maize which has ten pairs of chromosomes has ten linkage groups. It is assumed that a linear arrangement of the genes along the length of the same chromosome are linked together.

Linkage maps The greater is the distance between two genes, the greater is the chance that a chiasma will occur between their loci and higher is the percentage of crossing over between them. If, therefore, the percentage of crossing over between various genes are determined experimentally, the genes can be mapped in their order to the chromosome.

Linseed *Linum sativum*. Fam. Linaceae. An oil and fibre bearing plant.

Lipochondria *See* Golgi bodies.

Liquorice Fam. Papilionaceae. *Abrus precatorius*. Useful plant.

Lithotrophs These are the bacteria that obtain their energy from the oxidation of inorganic reduction *e.g.* sulphur and iron bacteria. These bacteria obtain energy as shown below:

$$2S + 3O_2 + 2H_2O \rightarrow 2H_2SO_4 + \text{energy}$$

and $$2H_2S + O_2 \rightarrow S_2 + 2H_2O + \text{energy}$$

Liverworts Bryophytes mostly with thalloid plant body. Bryophyta is divided into two classes *viz.* Hepaticae or liverworts and Musci or mosses.

Locomotion Plants usually do not show locomotion since they are rooted and fixed in the soil. But there are certain lower plants like *Chlamydomonas* and *Volvox,* which show locomotary movements while some animals like sea anemones, corals and sponges remain fixed to one place, on the substrate throughout most part of their life.

Lodicule In family Gramineae, the parianth is represented by two minute scales called lodicules.

Locus The fixed position of a gene in the chromosomes.

Lomentum When the pod is constricted or partitioned between the seeds into a number of one-seeded compartments.

Loofah *See* bath sponge.

Lophotrichus Such bacteria who have a tuft or flagella present at one end.

***Loranthus* sp.** A partial parasite of trees.

Lotus *Nelembo* sp. It is partially submerged plant.

Love thorn Fam. Gramineae. *Chrysopogon aciculatus.* It is a pasture land grass.

Lucerne Also called alfaalfa. Fam. Leguminoseae. A useful natural fertilizer used as green manure and also green fodder for animals.

Luffa spp. Fam. Cucurbitaceae. *Luffa acutangulis,* ribbed gourd. Bath sponge guard is *L. cylindrica.* Important vegetable plants.

Lupins Fam. Papilionaceae. *Lupinus* sp. Ornamental plant.

Lycopersicon *See* gametophytic incompatibility.

Lycopersicum esculentum Fam. Solanaceae. Tomato. Important vegetable and food plant.

Lyrate This is the shape of the leaf, when the shape of the leaf is like that of a lyre *i.e.* terminal lobe and some smaller lateral lobes.

Lysigenous During the development of mass of cells, their walls break down and dissolve, and as a consequence large irregular cavities appear; these are known as lysigenous cavities.

Lysigenous cavity The xylem has two metaxylem vessels, a protoxylem vessel, a few trachieds, parenchyma and the lysigenous cavity.

Lysis Disintegration or dislocation, usually the destruction of a bacterial host cell by infecting phage particles. *See* infection.

Lysogenic bacteria Living bacteria cells harbouring temperate phages (viruses).

Lysogeny Besides the phases, there is other type of phase called temperate phase. If this phase is introduced in a suspension of host cell, very few host cells show lysis, while most of them continue their normal growth.

Lysosomes Cytoplasmic organelles of animal cells which contain digestive enzymes for inracellular digestion of bacteria and other foreign bodies which the cells by the process of phagocytosis or pinacocytosis. They may cause cell destruction in ruptured condition.

Lythrus *See* tristyly.

Madder Fam. Rubiaceae. *Rubia cordifolia.* Medicinal plant.

Main gene The main or major gene gives expression to a character and the minor or modifying gene slightly alters the expression of a major gene but has not effect on the major gene.

Maintenance of male sterile line In an isolated field, the male sterile line A and the maintainer line B are grown. For every six rows of A line, two rows of B line are grown. The B line is identical to the A line except that it is male fertile. The B line has fertile cytoplasm but has the fertility restoring genes. The two lines are left for open pollination. Seeds are harvested separately from the A and B lines. In an isolated plot, the male sterile line A and the restorer line R are grown. For every six rows of A line is male fertile and possesses genes that restore male fertility. The two are left for open pollination. Seeds harvested from A line are the single cross hybrid seeds which are distributed to the farmers.

Maize *Zea mays,* family *Gramineae.* An important cultivated plant, used for food and fodder, and feed. A grain crop plant.

Male gamete As the pollen tube approached the ovule, the generative nucleus divided by mitosis (second pollen mitosis) to form two spermonuclei or male gametes. *See* life cycle of flowering plants.

Male gametophyte The pollen is the make gametophyte.

Male spores Also called microspores, are formed in anthers.

Male sterile line A line which has a condition in which pollen is absent or non functional in flowering plants.

Mallow Fam. Malvaceae. *Malva verticillata.* Green leaves used as vegetables.

Malvaceae A family of angiosperms. The family included about 80 genera and 1200 species. They are widely distributed in the temperate and tropical regions. The number of species increased as we reach tropics. A large number of cultivated plants are found in India. The floral formula is:

$$\oplus \; ⚥ \; \text{Epi } (6\text{–}7) \; K_5 \text{ or } (6) \; C_5 \; A \; \underline{G_5 \text{ or d}}$$

Mammoth tree Such a tree is tall as 90 m and over.

Mango Fam. Anacardiaceae. *Mangifera indica.* A famous fruit tree bearing the fruit is drupe and stone-fruit.

Mangrove Halophytes growing on sea coasts and estuaries, and also in salt marshes and salt lakes occasionally inundated by sea tides, form a special vegetátion called mangrove.

Manihot esculentus Topaco is the large fleshy root of *M. esculentus.* It is a perennial shrub grows in moist conditions.

Manometer An equipment used for measuring root pressure.

Map unit In mapping genes, a unit of distance must be used and this unit is called a map unit which is the space within which one percent of crossing over takes place.

***Maranta* sp.** Arrowroot has moniliform or beaded roots; is eaten as human food.

***Marchantia* sp.** It is a rosette type of thalloid liverwort (much larger than Riccia). Showing conspicuous dichotomous branching with a distinct midrib.

Margarine An imitation 'buffer' is made from the ground-nut oil as an industrial product.

Margosa *Azadirachta indica.* Used as timber and has medicinal value.

Mass flow In the mechanism of phloem transport, the carbohydrates move through the sieve tubes at the rare of about 100 cms per hour. Since this rate cannot be achieved simple dissuasion, it is commonly believed that the phloem transport occurs as mass flow of solutions from one end to the other. The mass flow or pressure flow means flow of solutions along the pressure gradients (from higher pressure to lower pressure).

Mass method of breeding Same as bulk method.

Mass selection A system of breeding in which seeds from individuals selected on the basis of phenotype is composited and used to grow in the next generation.

Master regulatory genes A small number of regulatory genes playing major role in establishing and maintaining the basic body plan in drosophila.

Mast tree *Polyalthia* sp. An ornamental tree.

Mather Kenneth Mather showed that the genetic components of a continuous variation are genes whose individual effects are not easily distinguished from environmental effects.

Mc clung The chromosome theory of sex determination was put forward by McClung, an American zoologist, who observed in 1902, that the male grasshopper possessed an odd number of chromosomes in contrast to the female which possessed an even number. The proof for this was furnished by Wilson and Steven who demonstrated in bugs that chromosome distribution followed a course exactly similar to that of sex distribution. This was a XX-XO condition.

Mc Fadden and Sears Kihara in Japan and Mc Fadden and Sears in USA succeeded in synthesising a hexaploid wheat by crossing the emmer wheat *(T. dococcoides)* with *Aegilops squarrosa* (2n-14) add then doubling the chromosomes of the hybrid.

Mechanical tissue Stems have to bear the weight of the upper parts. Thus the best position for strengthening tissues in stems, is close to the periphery, either in the form of a cylinder or in patches. Collenchyma and schlerenchyma including wood fibres and best fibres are the two most important tissues concerned in the strengthening of the plant body.

Mechanism of photosynthesis In the nineteenth century it was believed that the raw materials essential condition and the end products whereas in the various types of reactions are taking place and different types of molecules are being formed in between. In twentieth century, great success has been achieved in understanding the process and discovering various intermediate components. The isotopic oxygen (O^{10}) has helped to know the source of oxygen and its fate in photosynthesis. When a plant was fed with water having isotopic oxygen (radiotracer method), the oxygen liberated was found to be isotope.

$$6CO_2 + 6H_2O \xrightarrow[\text{Chlorophyll}]{\text{Light}} C_6H_{12}O_6 + 6O_2$$

water prepared with isotopic oxygen.

Mechanism of respiration Respiration as a whole consists of three inter related processes as:

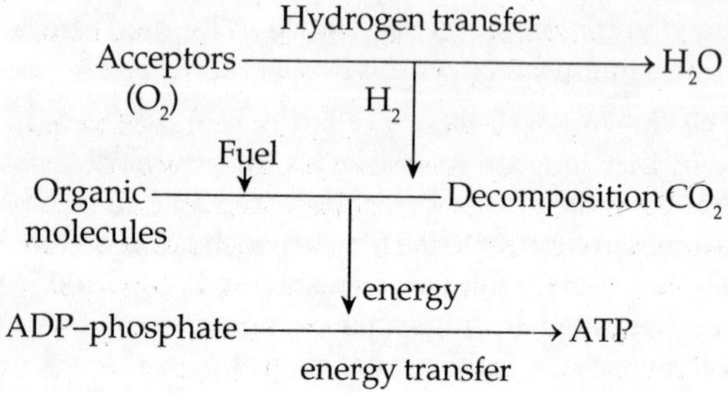

The three phase pattern of respiration

Medicago sativa Fam. Papilionaceae. *See* lucerne.

Medicinal plants Our forests abound in medicinal herbs, shrubs and trees, yielding many valuable drugs. These are plants which are used in curing diseases. The old Indian scriptures have devoted space to collection and preparation of drugs from these plants successfully.

Medulla It forms the central core of the stem and the root and is usually made up of large celled parenchyma and with abundant intercellular spaces. It is also known as pith.

Medullary rays Each extension of the pith towards the pericycle which is a strip of parenchyma, is called the medullary ray or pith ray. In vascular bundle, the pith extends in between the adjacent vascular bundles in the form of pith rays or medullary rays.

Megasporangium In Gymnosperms, each ovule consists of a mass of parenchymatous cells called megasporangium or nucellus.

Megaspore In higher plants, the megaspore mother cell divides in four megaspores of which three degenerate and the fourth develops into the embryo sac. The microscope mother cell divides into a tetrad of microspores or four pollen grains.

Megaspore mother cell In angiosperms, in the nucellus, one of the cells gets differentiated and forms the embryo sac mother cell or megaspore mother cell.

Megasporocyte The female generative body producing *female* spores or embryo mother cells (E.M.C.) located with the ovules.

Meiocyte Any cell that undergoes meiosis.

Meiosis Two successive nuclear divisions in the course of which the diploid chromosome number is reduced to the haploid. Meiosis formation of four nuclei, each with one chromatid (*i.e.* one daughter chromosome) from each bivalent (*i.e.* two homologous chromosomes). It thus brings about a reduction in the number of chromosomes from the diploid to haploid state.

Meiospore In plants, one of the asexual reproductive cells produced by meiosis from a meiocyte.

Meiotic drive Refers to preferential segregation of the particular chromosome into the games in every meiotic division.

Melon Fam. Cucurbitaceae. *Cucumis melo.* Musk melon. The water melon is *Citrullus lanatus,* both of these species bear sweet edible fruits.

Mendel it was in 1865 when Gregor Johann Mendel presented a paper embodying the results of his experiments on hybridization in peas.

Mendel's experiments His series of experiments involved seven pairs of contrasting characters of the edible garden peas, *Pisum sativum.* These characters were–smooth and wrinkled ripe seeds; yellow and green peas, tall and short stems; axial and terminal flowers; white or grey seed coats; yellow and green cotyledons etc. Peas are normally self-pollinating but Mendel tried the experiment of transferring the pollen of one variety to the stigma of other. He first crossed plants differing in a single pair of contrasting characters such as tallness and shortness of plants. The flowers of F_1 generation were allowed to self-pollinate. All the resultant seeds were collected and sown to give the second filial or F_2 generation. In this generation Mendel found tall plants and short plants in the proportion of 3 tall and 1 short. Finally all the F_2 generation were obtained. Seeds from the short plants of the F_2 generation gave rise to nothing but the talls in the F_3 generation. But the talls of the F_2 did not behave in the same way. One third of them gave rise to nothing but talls in the F_3 generation but the other two-thirds segregated like the plant of the F_1 generation, giving rise to short and tall plants in the proportion of 3 : 1

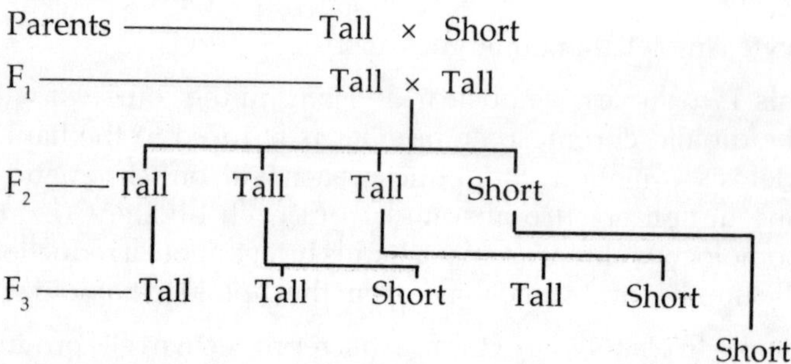

Mendel's law The first law is that hereditary factors (genes) are formed in pairs in mature individuals. They do not blend but separate or segregate unchanged during the formation of gametes. The second law is that the members of different pairs of factors responsible for different characters segregate and recombine independently in different gametes.

Mendel's dihybrid ratio Dihybrid means a cross involving two characters e.g. (1) shape of the seed and (2) colour of the seed. The two pairs of alleles for shape are (*a*) round and (*b*) wrinkled and for the colour, the two alleles are (*i*) yellow and (*ii*) green. A cross involving the two characters gives the F_1 as Round-Yellow. But the F_2 segregates in 9 round and yellow; 3 round green; 3 wrinkled yellow and 1 wrinkled green.

Mendel's principles of heredity *See* Mendel's laws.

Mendel's second law Law of segregation.

Mendel's laws' cytological evidence The independent assortment of members of different pairs of alleles according to the second law of Mendel is possible only because the members of various homologous pairs of chromosomes are distributed to the gametes independently of each others. The homologues are in pairs and so exist the two alleles of a character in pairs.

Mendelism Gregor Johan Mendel formulated the law of inheritance. This knowledge of inheritance is known as Mendelism as he was a pioneer in this field.

Mericlinal chimaera *See* graft chimaera.

Meristele In Pteridophytes, the ground tissue reach directly upto the epidermis. Hypodermis is followed by parenchymatus ground tissue in which are present two rings of vascular strands or meristeles.

Meristem A cellular region in plants characterised by repeated cell division and wherein, therefore, growth in size originates. The cells of meristematic tissue are small in size, spherical, ovalor polygonalin shape and are almost isodiametric. The cells are closely packed without an intercellular space. They are densely packed with active protoplasm, without vacuoles or with very small vacuoles and with a distinct nucleus. The cell walls are thin and homogenous. They are active state of division or have retained this power.

Meristic traits (characters) Traits in which the phenotype is determined by counting.

Merozygote Parental zygote produced by a process of partial genetic exchange such as transformation in bacteria.

Mesarch In certain cases, the differentiation starts the middle of the procambium and consequently the protoxylem becomes surrounded by the metaxylem. This condition is known mesarch and is found in ferns.

Mesocarp *See* endocarp and cremocarp.

Mesogamy In angiosperm, the pollen tube enters the ovule *i.e.* nucellus to reach embryo sac. The pollen tube may enter the ovule through micropyle when it is known as porogamy or through chalazal when it is known as chalazogamy or through any other place when it is known as mesogamy.

Mesophyll The ground lying between the upper epidermis and the lower one. It is differentiated into (*a*) palisade parenchyma and (*b*) spongy parenchyma.

Mesophytes Such plant ecologically lying between the hydrophytes and xerophytes.

Messanger RNA Ribonucleic acid transcribed by DNA and conferring amino acid specifically on the ribosome.

Metabolic stage A cell which is not in the process of division is said to be interphase (or interkinesis). It is often called the resting stage because there is no dividing activity in the cell at this time. This is also called the metabolic stage.

Metacentric The metacentric chromosomes are V-shaped and in these the centromere occurs in the centre and forming two equal arms. The amphibians have metacentric chromosomes.

Metakinesis Refers to the movement of the chromosomes towards the central region of the cell just prior of metaphase.

Metaphase The stage of mitosis or meiosis during which the chromosomes lie on the equatorial plane of the spindle.

Metaphloem The outer portion of the phloem, which is a broken mass, is the protophloem and the kinner portion is the metaphloem.

Metaxylem The two smaller vessels (annular and spiral) lying radially toward the centre constitute the protoxylem, and the two bigger vessels (pitted) lying laterally together with the small pitted trachieds, lying in between them constitute the metaxylem.

Metazoa Keeping all the knowledge available so far, recognises four basic categories or organisms instead of two. These categories are named as Monera, Protista, Metaphyta and Metazoa.

Microcentre Areas within centres of origin that show greater diversity than the remaining centres of origin.

Microgametes In Algae, in anisogamous forms like *Chlamydomonus braunii*, the female cell produces four large microgametes and the male cell 8 smaller microgametes. In anisogamous form if Algae like *Chlamydomonas braunii*, the female cell produces four large microgametes and the male cell of microgametes.

Micron 0.001 millimeter.

Micropyle Just below the hilum a very minute slit is seen. This opening is the micropyle.

Microsomes These produce cell proteins.

Microsporangia These are the pollen chambers.

Microspores The pollen grains.

Microspore mother cells Pollen mother cells.

Microsporophyll A male cone in Gymnosperms, has a central axis on which a larger number of microsporophylls or stamens are spirally arranged. Each microsporophyll bears two microsporangia or pollen sacs in its abaxial (lower) surface, one on either side of the median line.

Middle lamella It is the first formed cementing wall between cells made up of pectin. The original or middle wall between two contiguous cells.

Millets These are smaller grain yielding plants, the grain being used as food and stem and leaves used as animal fodder.

Mimicry *See Arisaema* sp.

Mimordica charantia Fam. Cucurbitaceae. Bitter gourd. A vegetable.

Mimoseae A family of angiosperms.

Mimosodeae The subfamily of Lelgumonaceae. Floral formula is:

$$\oplus ♂ K_5\ C_5 \text{ or } (5)\ A_2 \text{ or } A\ (2)\ \underline{G1}$$

Mineral crystals These occur either in the cell cavity or the cell wall and are mostly of silica, calcium carbonate and calcium oxalate.

Mineral nutrition The fact that plants yield an ash containing inorganic salts, and the beneficial effect on crops of adding this ash to the soil was known since antiquity. *See* Soil Science.

Minor genes Same as modifying genes.

Mint Fam. Labiatae. Mentha viridis. A medicinal plant.

Mirablis jalapa With his studies on this plant Mendel provided that a single pollen grain fertilises the single egg.

***Miriabilis* sp.** Four O'clock plant, fruits are achenes.

Mistletoe Viscum sp. It is a parasite of other plants.

Mitochondria These are bodies present in cells which are centres of enzymatic activity.

Mitochondrion It is also called the power-house of cell. Particularly their membrane are a fine example of nature's combination of design and function. They possesses a structure that provides a most efficient flow of energy in the form of ATP molecules.

Mitogenesis Induction of mitosis.

Mitosis A process of nuclear division in which chromosomes divide longitudinally and two daughter nuclei are formed, each having a chromosome complement equal to that of the original nucleus.

Mitotic index The number of cells undergoing mitosis per thousand cells.

Mitotic inhibitor A compound that suppresses mitosis.

Mitotic poison A compound that prevents affects of the complete mitosis.

Modification of leaf These are leaf tendrils, leaf spines, phyllods etc.

Modification of roots Sometimes the roots have to perform certain special functions like storage of food, giving support to plant etc. For this purpose, they assume special form and shape and are known as modified roots. They may be A. fleshy roots B. nodulated roots C. These are all tap roots modifications. Modifications of adventitious roots are A. Fleshy roots, B. Fasciculated roots, C. Palmate, D. Tuberous, E. Nodulose, F. Annulated, G. Beaded or Moniliform. For mechanical support– A. Prop roots, B. Still roots, C. Climbing or clinging roots and for various vital functions–A. Parasitic roots, B. Respiratory roots or breathing roots or Pneumatophores, C. Assimilatory roots, D. Epiphytic roots.

Modification of stems As far as texture is concerned, it can be woody or solid or hollow (fistular) and in shape, it can be circular or rounded or angular. As far the surface is concerned, it can be glaucous, glabrous, prickly, hairy. The form of stems can be Aerial, 2. Erect, 3. Excurrent, 4. Deliquiscent, 5. Caudex, 6. Culm. The weak stems can be trailers, twiners climbers. The underground stems are–1. Tuber, 2. Rhizome, 3. Corm, 4. Bulb, 5. Tunicated, 6. Scaly bulb. The subaerial modifications are runner, sucker, offsett, stolon. Aerial modification are tendrils, spines, phylloclade.

Modifying genes Sometimes a certain character is determined fundamentally by one gene but is modified slightly by other genes. The main or major gene gives expression to a character and the minor or modifying genes slightly alter the degree of expression of this character.

Moll's experiment It is an experiment to demonstrate that plants cannot photosynthesise unless carbon dioxide is available. This experiment proves that carbon dioxide is essential for the photosynthesis.

Monadelphous stamens When all the filaments are united together into a single bundle but the anthers are free, the stamens are said to be monadelphous.

Moniliform root *See* beaded roots.

Monk's hood Fam. Ranunculaceae. *Aconitum ferox*. Medicinal.

Monochasial cyme In this type of inflorescence the main axis ends in a flower and at the same time, it produces only one lateral branch at the time ending in a flower. This is also called Uniparous cyme.

Monocotyledons A class of angiospersm.

Monoecious Unisexual or diclinous flowers *i.e.* separate male and female flowers may be borne by one and the same plant. Individuals producing both sperms and eggs.

Monohaploids These are haploids which arise from true diploids and whose chromosomes are therefore non-homologous to one another *e.g.* haploids of *Zea mays*.

Monohybrid ratio A cross between parents differing in a single gene. The phenotypes appear in a 3 : 1 ratio. When only one pair of characters *i.e.* red colour and white colour of flower is taken into consideration, the individual formed on crossing will possess mixed genes *i.e.* both the alleles and are called monohybrids and F_2 ratio of 1 : 2 : 1.

Monomorphic locus A locus at which the most common allele has a frequency of greater than 0.95.

Monoploid An organism with a basic number of chromosome.

Monoploid number Haploid or basic number.

Monopodial When the growth of the main stem is indefinite *i.e.* it continues to grow indefinitely by the terminal bud and give off branches laterally in acropetal succession *i.e.* the lower branches are older and longer than the upper ones. This is a recemose type of branching.

Monosome An organism lacking one chromosome of the diploid complement, hence having 2n-1 chromosomes.

Monosomic A monosomic lacks, one chromosome of the normal complement of the somatic cells (2n-1).

Monotrichous These bacteria have a single flagellum present at one end.

***Monotropa* sp.** This is a seprophytic plant.

Monsoon forest The tall growing extensive forest of tropical Asia that experiences a marked hot dry season. Monsoon forests resemble the equatorial rain forests, the main difference is that the monsoon forests can have also such flora which have deciduous trees also.

Moon flower Fam. Convolvulaceae. *Ipomea grandiflora.* Ornamental.

Morgan Thomas Hunt Morgan found that some 'factors' in the fruit fly Drosophila were not inherited at random but in groups. He along with his associates, Muller, Stertevant and Bridges, discovered that the number of such groups in Drosophila was four, corresponding to the number of chromosomes in the germ cells.

Morgan's theory That specific genes are borne in specific chromosomes. Study of linkage and crossing over in *Drosophila melanogaster* by Morgan, and associates threw more light on the one hand and the material chromosome on the other.

Morning glory Fam. Convulvulaceae. *Ipomea purpurea.* Ornamental.

Moss There are Bryophyta with leafy stem.

Movements These are the sign of life but most plants are fixed to the ground. The capacity of plants or their particular organ receive stimuli from outside and to respond to them is called irritability, which in one way or another expresses itself in some kind of movement. These are commonly recognised as a characteristic of a living organism. Plants posses restricted movements while animals have conspicuous one.

Mucilage It is a slimy substance, which absorbs water freely and retain it.

Mucilagenous canal In Gymnosperms, the cells of the cortex are densely filled with starch grains. A number of mucilage canals are found in the cortex.

Musci *See* Liverworts.

Musilageination Sometime the cellulose gets converted into a slimy substance called mucilage.

Mucuna puriens Fam. Papilionaceae. Cowage.

Muller Discovery of Muller that X-rays could be used to induce mutations opened a new line of work.

Multiple alleles A series of alleles each differing from the other but all alternatively situated at the same chromosomal locus. Such alleles forming a series of genes that are alternative to one another in inheritance because they are situated at the same locus of homologous chromosomes.

Multiple allelic genes Different mutations at the same locus give rise to multiple allelic series.

Multiple factor Quantitative characters are governed by many genes. Such genes are called as multiple genes, or multiple factors or polygenes or polymeric genes. An inheritance understanding of characters like length of ear in corn, yield of grains in rice, yield of eggs in hens, yield of milk in dairy cattle, is not so easy because the individuals show subtle differences and can be separated into classes only after making measurements. These characters are examples of quantitative character that show more or less continuous variations and are governed by a large number of genes called multiple genes, or multiple factors of polymeric genes or polygenes.

Multiple factor hypothesis Nilsson Ehle published his multiple factor hypothesis.

Multiple fruits Such fruits develop from such an inflorescence where the flowers are crowded together and often fused with one another. These are also called composite fruits.

Multiple genes Three or more pairs of genes with a cumulative effect. Same as multiple factors. Quantitative characters are controlled by the joint action of a very large number of multiple genes which cannot be distinguished from one another because their individual effect on the phenotype are insignificant in comparison to the fluctuations due to the environment.

Multivalent Association of more than two homologous chromosomes at meiosis.

Muntzing Many different diploid varieties have been made into tetraploid by Muntzing such as in rye (2n-14) made in tetraploid 2n-28.

Mushrooms *See Agaricus* sp.

***Mussenda* sp.** Fam. Rubiaceae. Ornamental.

Mustard Fam. Cruciferae, rape seed. Important edible oil bearing plant.

Mutagen Attempts to induce mutations artificially were not successful till 1927 H.J. Muller discovered the magnetic action of X-rays on Drosophila.

Mutant An individual which has suddenly acquired a heritable variation not present in the parent form.

Mutation A sudden heritable variation. A gene mutation is a change in a gene form, one allelic form to another.

Mutation breeding The mutation frequency of genes can be increased to a good extent by irradiation or chemicals. It is also possible to create mutants artificially, but because of the low frequency of viable mutants which are really beneficial, their detection and isolation is a difficult task.

Mutation theory *See* de Vries.

Muton A smallest segment of DNA or a subunit of a cistron that can be changed and thereby bring about a mutation.

Mycorrhiza In some cases as in certain Gymnosperms, absorption is helped by mycorrhiza which is an association of fungus with the roots.

n and 2n The gametic and zygotic (somatic) chromosome numbers respectively of a species.

Napiform root When the root is considerably swollen at the upper part becoming almost spherical and sharply tapering at the lower part.

***Naravelia* sp.** Fam. Rununculaceae. A climbing shrub.

Nastertium indicum Fam. Cricifereae. Ornamental plant.

Nasties Nastic movements. These are the movements of mainly dorsiventral organs like leaves or petals caused by extreme stimuli.

Natural polyploids Polyploids arise in nature most commonly by failure of meiosis that result in the formation of unreduced gametes. They may also be formed from somatic cells in which a failure of mitosis has resulted in doubling of the chromosome complement. Cultivated banana and tobacco are examples of natural polyploids.

Natural selection The theory of natural selection was put forth by Charles Darwin in 1858 and later by Wallace independently. Darwin assumed that hereditary particles formed pangenes or gennules, are produced by every part of the body during the life time of an organism and that, these assume the characters of the various parts of the body from which they are derived, together with whatever modifications the latter have acquired. Eventually all pangenes accumulated to form germ cells which give rise to the new individuals, thus ensuring the development of the parental characters and inheritance of acquired characters. Also many more individuals and consequently there is always a struggle for existence. If hereditary differences occur within the wild species of plants, nature will eliminate some and select others.

Natural vegetation Vegetation that has developed in an environment untouched by man and that therefore closely corresponds to the prevailing conditions of soil and climate.

Nebel Nebel and Blakeslee found that the alkaloid colchicine derived from the corms and the seeds of the *Sutumn crocus (Colchicum autumnale)* of family Liliaceae, was very effecting in doubling the chromosome number, and easy and effective method for production of polyploids. It was an important tool available to the plant breeders.

Neck canal cells In Bryophyta, in the cavity of the neck, there are usually four neck canal cells.

Nedulose root When the slender root becomes suddenly swollen at or near the apex.

Needle The leaves of pines are acicular in outline and are known as needles.

Ne-penthes Pitcher plant. It is a carnivorous plant.

Nerium indicum Fam. Apocynaceae. Yellow oleander. Medicinal.

Nettles These plants bear glandular hairs secreting, irritating poisonous substances.

Neurospora *See Escherichia coli.*

Neutral alleles Same as inert alleles.

Neutrons An atom is made up of two parts (1) an inner positively changed nucleus consisting of positively charged particles called protons and (2) an outer constellation of negatively charged electrons. The charges are so balanced that an atom is electrically neutral.

Nicker bean Fam. Mimosae. *Entada gigas.*

Nicotiana *Nicotiana* is a plant in which lot of studies have been conducted in respect to the series of alleles that determine incompatibility among species. It is a good material for the purpose of genetical studies.

Nicotiana tabaccum Fam. Solanaceae. Tobacco plant. Narcotic.

Nightshade Fam. Solanaceae. *Atropa bellodanna.* Medicinal plant.

Nigella sativa Fam. Rununculaceae. Black cumin. Medicinal use.

Nilsson Ehle *See* multiple factors.

Nitrogen cycle The plants absorb nitrates from the soil and synthesise proteins by adding the nitrogen from nitrates to carbon, hydrogen and oxygen. Protein in their turn are used by the plants or the animals in forming protoplasm. Hence nitrogen is very essential for life but the green plants cannot utilise the atmospheric nitrogen unless and until it is converted into nitrates or ammonia (NH_3).

Nitrogen fixation *See* Soil Science.

Nobilization The process of the improvement of the hybrid by back crossing with the noble canes is called 'Nobilization'.

Nodes and internodes These are always present in the stem but absent in roots.

Non-cross over gametes These are the parental gametes, other than cross over gametes.

Non-cyclic electron transfer In this sequence, high energy electrons from chlorophyll are again believed to be captured first by ferrodoxin from where they are continued to be transferred to NADP (Nicotinamide Adenine Denucleotide Phosphate) which becomes electrically negatively charged. NADP is a hydrogen carrying coenzyme which in photolysis comes to carry the hydrogen released as a result of the splitting of water.

Non-disjunction *See* disjunction.

Non-heritable variations These are variations due to environment.

Non-ionising radiations Such radiations are less drastic than the ejection of electrons but do result in the increased activity of the affected molecules leading to mutations.

Non-parental gametes These are the crossover gametes.

Non-recurrent parent Donor parent; that parent of a hybrid which is not utilised as a parent in back crossing.

Non-recurring parent Donor parent; it is that parent which is not again utilised as a parent in back crossing.

Nuclear membrane The thin wall surrounding the nucleus.

Nuclear sap The fluid material enclosed in the nucleus.

Nuclei *See* sperm nuclei, generative nuclei etc.

Nuclei acid DNA or RNA.

Nucleocide Portion of a DNA or RNA molecule compound of one deoxyribose molecule (in DNA) or ribose in (RNA) plus a purine or a pyrimidine.

Nucleoli Lying in the nucleus are one or more rounded bodies called nucleoli (sing. nucleolus).

Nucleotide Portion of DNA or RNA molecules composed of one deoxyribose phosphate unit (in DNA) on ribose phosphate unit (in RNA) plus a purine or pyrimidine.

Nucleus seed Breeder's seed produced by the sponsoring plant breeder or institution and which provides the source for the initial and recurring increase of foundation seed.

Nullisomics An individual lacking one pair of chromosomes of the diploid complement, hence with 2n-2 chromosomes.

Nutation When the stem tip moves from one side to another in a zigzag manner, this movement is nutation.

Nutrition The living organisms can take in simpler substances from the outside world and convert them into complex substances of which their bodies are made.

Nux-vomica *Strychnos* sp. Medicinal importance.

***Nyctanthes* sp.** The flowers of this plant give scented flowers which give sweet in night. Also called queen of night.

Nyctinastic Floral parts and leaves take different positions by day and night. Such diurnal movements are known as nyctinastic movements.

Nyctinasty It is the movement induced by alternation of day and night.

***Nymphaea* sp.** Partly submerged plant. Water lily.

Oak *Quercus* sp. A good fuel wood tree. Fruit is a nut.

Oat Fam. Gramineae. *Avena sterilis.* A hardy cereal that grows in cooler climate than wheat or barley. It is mainly used as fodder for animals and is also used as human food.

Object of backcrossing The backcross in a form of a recurrent hybridization by which one or two desirable characteristics are added on to an otherwise superior variety.

Object of hybridization Is to combine in a singular variety the desirable characters of two or more lines, varieties or species.

Obligate anaerobes *See* Plant Pathology.

Observation plot In the first step, the previously selected plants are put as single plant rows in the observation plots, where preliminary observations are recorded.

Ochikers Ochikers in 1943 showed that a mixture of ethylurethane and potassium. chloride induced tanslocation in *Oenothera.*

Ochoa Koerberg and Ochoa shared the Nobel Prize in 1959 for elucidation of DNA.

Ochreate This kind of stipules form a hollow tube encircling the stem from the node upto a certain height of the internode in front of the petiole.

Ocimum sanctum Fam. Labiatae. Sacred basil. An essential oil bearing plant. Medicinal plant.

Octant In the ovule of the angiosperm, the embryonal cell enlarges and divides by three successive walls at right angles to one another, the first one being transverse and gives rise to an eight celled structure called octant. Four of the cells of the octant lie towards suspensor and are known as posterior octant and those on the other side are called anterior octants.

Oenothera De Vries observed that the evening primerose *Oenothera lamarkiana*, a native of America, was growing in Holland. In a population of this weed, he observed some plants which differed in some characters from the typical *Oenothera lamarkiana*. Since it is a self fertilized species, he felt that these variants have risen suddenly rather than as hybrids. He observed that variants continued to appear in his plantings for several years. He also saw that these variations were inherited. He called these drastic changes as mutations and maintained that mutations play an important role in the evolution of new species.

Oenothera lamarkiana Evening premrose. De Vries' plant of mutations observation.

Offset Like the runner, this originates in the axil of a leaf as a short, more of less thickened, horizontal branch.

Off type In selection procedure, only those plants are selected which conform to the requirement of breeder, others which are not required any further are the off types, to be discarded from any further use in the breeding programme.

Oligogenes Major genes controlling qualitative characters which show normal mendiliean inheritance.

Olive The fruit of an evergreen tree that grows in Mediterranean region. The fruit is used for pickle or also canned. Most of the fruit is pressed to yield a rich oil which can be used for cooking or as an alternative for butter.

Onion *See Allium cepa.*

Ontogeny The complete development of the individual from zygote, spore etc. to adult form.

Oocyte The egg mother cell.

Oogenesis The process of development and differentiation of the ovum from the oogonium.

Oospore In Bryophyta, the cavity of the ventre contains a large almost spherical egg or oosphere.

Operator site A segment of DNA in an operon that affects activity or non-activity of associated cistrons; it may be combined with a repressor and thereby 'turn off' the associated cistrons.

Operculina turpethum Fam. Convolvulaceae. Indian Jalap. Medicinal plant.

Operculum In the mass capsule, it is the circular cup shaped on the top of capsule.

Operon A system of cistrons, operator and promotor sites, by which a given genetically controlled metabolic activity is regulated.

Oppositional factor hypothesis In *Nicotiana sanderae*, pollen tubes are unable to grow down a style if both the pollen tubes and the style has the same allele. Three main kinds of pollinations were observed (1) fully incompatible, (2) half the pollen incompatible, (3) fully compatible.

***Opuntia* sp.** The prickly pear. A xerophytic species.

Orbicular Or rotund, when the blade is more or less circular in outline.

Orchids The epiphytic roots of orchids sometime turn green. These are distinctive plants and are highly in the international market as in florist trade due to their intricately designed, spectacular flowers, brilliant colours, delightful appearances, myriads of sizes, shapes, forms and long lasting qualities. These belong to most highly family of *Orchideaceae* of monocotyledons.

Orientation The movement of chromosomes so that their centromere lie axially with respect to the spindle.

Orthostichy The vertical row of leaves in leaf phyllotaxy.

Orthotropous When the ovule is erect or straight so that the funicle, chalaza and micropyl lie on the same vertical line.

Oryza sativa The rice plant has been subject to improvement by plant breeding methods because it is an important staple crop of majority of world population. Genic sterility in individuals in which there are completely normal pairing of chromosomes is usually due to gene. This kind of sterility was observed in hybrids between races *indica* and *joponica* of *Oryza sativa*.

Oscillatoria A dark green filamentous algae.

Osmosis The process by which osmotic pressure is created is called osmosis.

Osmotic pressure The protoplasm of a cell is rather concentrated solution (*i.e.* water molecules are relatively few), the cell membrane is differentially absorbent and the environmental solution is usually less concentrated (*i.e.* water molecules are more abundant per unit volume) than the protoplasm. Water now moves into the cell diluting the cell contents and increasing the cell volume. If the cell wall exerts a resistance to the volume increase caused by the water uptake, a pressure will be developed on the cell contents in this way is called osmotic pressure.

Otocephaly Abnormal development of the head of a mammalian foetus.

Outcross A cross usually natural to a plant with different genotype.

Ovary The female gonad in animals or the ovule containing portion of the pistil of a flower.

Ovary transplantation experiments The independence of the germplasm from the somatoplasm was shown by ovary transplantation experiment of Castle and Philips.

Ovate When the blade is egg shaped *i.e.* broader at the base than at apex.

Ovule Cycus ovules are the largest in the world. The ovules are orthotropous and sessile. The basic structure of ovules in all angiosperms is the same. The main body of the ovule consists of mass of parenchymatus cells as nucellus, has funicle, hilum, raphae and chalaza.

Ovuliferous scale In Pinus, on upper surface of each bract scale, there is another stouter and bigger scale, woody in nature and somewhat triangular in shape. The second scale of the pair is large stout and woody and is known as ovuliferous scale.

Ovum The female gamete, egg.

***Oxalis* sp.** A common weed, the wood sorrel. Has a runner stem.

Oxylophyte A plant that thrives in or is restricted to acid soils.

P_1 P_1 are the parents of F_1 generation. P_2, the grandparents. Also used to designate different parents used in making hybrids.

Pachynema Also called pachytene. The chromosomes become shorter and thicker as synapsis continues.

Paederia foetida Fam. Rubiaceae. Medicinal plant.

Pagoda tree Fam. Apocynaceae. Also called Temple tree, *Plumeria rubra.*

Painter The discovery by T.S. Painter of the usefulness in cytogenetic research of the giant chromosomes in the salivary gland cells of the Drosophila provided an additional avenue of approach to understanding of the chromosomes.

Paired plots An experimental field design in which treatments are tested and are paired in such a way that no one treatment is always placed to only one side of the other.

Palea In a spikelet, opposite to lemma with a flower in its axil and opposite to lemma there is a small two-nerved bracteole called palea.

Palindrome A segment of DNA in which the base pair sequence reads the same in both directions from a point of symmetry.

Palisade parenchyma These are usually one, two or two or three layers of elongated, more or less cylindrical cells packed with their long axes at right angles to the epidermis in the leaf.

Palma Family Palmae. Economically this is one of the most important families. Many palms such as Palmyra palm *(Borassus flabellifer)*, toddy palm *(Caryota urens)*, date palm *(Phoenix sylvestris)*, double coconut palm *(Lodoicea maldivica)*, talipot palm *(Corypha umbraculifera)*, oil palm *(Eaeis gineensis)*, coconut palm *(Metroxylon rumphil)*, nut palm *(Area catechu)*, sago palm *(Metroxylon rumphil)*, can *(Calamus* sp.*)* are the important palms.

Palmae A family of angiosperms, monocotyledons.

Palmate Palmately compound leaf is one in which the petiole bears terminally, articulated to it, a number of leaflets which seem to be radiating from a common point like fingers from the palm.

Palmate venation In a palmately veined leaf the incision of the leaf blade it passes towards the base of the mid rib, while in case of pinnately veined leaf the incision of the leaf blade proceeds from the margin towards the midrib.

Palmella In *Chlamydomonas* (Algae), this method of asexual reproduction is very common under unfavourable conditions. It is characterised by the failure of flagella formation by the daughter protoplasts and gelatinization of the cell membrane of the daughter protoplasts.

***Pancreatium* sp.** Fam. Amaryllidaceae. Spider lily.

***Pandanus* sp.** A xerophyte.

Pangenes *See* natural selection.

Pangenesis hypothesis After the publication of origin of species, Darwin adopted the doctrine of the inheritance of acquired characters but he proposed a new theory of how it happened. He modified the views of Spencer and proposed the hypothesis of Pangenesis.

Panicle When the main axis of the receme is branched and the lateral branches bear the flowers, the inflorescence is said to be compound race of panicle.

***Papaver* sp.** Fam. Papaveraceae. Poppy. Yields morphine from unripe fruits.

Papaw Fam. Cariaccai. *Carica papaya.* It is a good source of papa in enzyme, bears excellent edible fruits.

Paplionaceae A subfamily of Leguminoseae. Includes many crop plants as pulse group and vegetables. This subfamily is the largest of the three subfamilies of Leguminoseae and is represented in India by about 754 species, which are mostly confined to eastern Himalayas and Assam. The floral formula is:

$$+ \male K_{(5)} C_{1+2+(2)} A_{1+(9)} \underline{G1}$$

Papilionaceous This type of flower is zygomorphic and polypetlous and is called butterfly-like. Such a flower is composed of five petals, of which outermost is standard or vexillum, the two lateral ones are known as wings or alae and the two innermost ones are the keel or carina.

Pappus A modification of calyx as in the sunflower family (Compositeae).

Paracentric inversion An inverted chromosome is one in which a portion of the gene sequence has been rearranged in a reverse order. If the rearrangement is confined to the single arm of chromosome, it is called a paracentric inversion.

Paradesmos The two basal granules in the neuromotor apparatus of Algae.

Parachute mechanism The pappus is persistent in the fruit and opens in an umbrellalike fashion, thus acting as a parachute to help the fruit to be carried by air current to distance.

Paramaecium It is said that inheritance of certain characters (Killer character) in *Paramaecium* is partly controlled by plasmogenes. The killer *paramaecium* contains a gene 'K' in the nucleus and also particles called Kappa particles, in the cytoplasm. The particles are made of DNA and secrete a substance called paramaecin.

Paramutation A mutation in which one member of a heterozygous pair of alleles under the influence of mutated allele changed its expression which is transferred to subsequent generation.

Paraphyses Many of the cells of the hymenium (rusts) bear spores and are called basidia; while others are sterile and are called paraphyses.

Paraphysis The paraphysis (Bryophyta) is a thin threadlike filamentous multicellular structure. The cells are arranged in a single row and the terminal cell of the filament is larger than the rest.

Parasites Plants that grow on other living plants.

Paratonic The movements shown by plant organs in an induced manner while the autonomous movement in spontaneous.

Parenchyma Consists of collection of cells which are more or less isodiametric *i.e.* equally expanded on all sides. Their wall are thin and of cellulose.

Parental gametes *See* non cross over gametes.

Parietal This is a placentation in which the ovary is one chambered and the placentae bearing the ovules develop in the inner wall of the ovary corresponding to the confluent margins of carpels.

Paripinnate *See* imparopinnate.

Parkinsonea aculeata Fam. Mimoseae. Jerusalem thorn.

Parsnip Fam. Compositeae. *Pastinica sativa.*

Parthenogenesis The development of the zygote from the egg-cell without the set of fertilization. In Algea *(Ullathrix)*, in some cases, the gametes fail to fuse and behave like zygospores. They withdraw their cilia, round off and secrete a thick wall and perenate through a unfavourable period. Such spores are called azygospores which germinate either directly or to form a green filament or may first produce aplanospores. This method of reproduction is called parthenogenesis.

Parthenocarpy The production of fruits without fertilization and normally without seeds.

Parthenospores Same azygospores.

Passion flower Papilionaceae. *Passiflora* sp. Ornamental.

Passive absorption The transpiring shoots can absorb water through anaesthecized or dead roots or even the absence of roots in case of cut shoots. This type of absorption is termed as passive absorption.

Pathogenecity variations Erikson in 1894 discovered that within a species of the pathogen, there may be several sub-species which differ from each other in their ability to attack only certain species and not others.

Pearl millet *See Pennisetum typhoides.*

Pedate When the leaf is divided into a number of lobes, which spread out like the claw of a bird.

Pedicel The stalk of the individual flower. The flower's long or short stalk.

Pedigree A record of history of the ancestry of an individual strain and variety.

Pedigree method of breeding In a self pollinated crop, the parents selected for hybridization will be homozygous of all loci. All the F_1 plants will look exactly alike and will have the same genotype, although they will be highly heterozygous. The hybrids will segregate for a large number of genes and in the F_2 generation, all the plants may differ from one another. Individual plants with the desired combinations of characters are therefore selected in the F_2 generation. With each succeeding generation, heterozygosity will be reduced by one-half. In the F_3 and F_4 generations, the plants would have become homozygous for several loci.

***Pedillanthus* sp.** Jew's slipper. Bears cythium type of inflorescence.

Pelican flower *Aristolichia* sp. Bears duck-shaped flowers.

Pellew *See* Keeble and Pellew.

Pennisetum typhoides Fam. Gramineae. *Bajra* or *Cumbu*. Pearl millet. It is this millet which has been fully exploited for production of hybrid in a cross pollination crop. Important food and fodder plant.

Pentamerous A symmetrical flower of which the number of parts in each whorl is 5 or any multiple of five. Trimarous, when the number of parts in each whorl is 3 or any multiple of three.

Pentastichous This is also called 2/5 phyllotaxy of leaves.

Pepo This is fleshy, many seeded fruit like the berry but it develops from an inferior, one celled or spuriously three celled, syncarpous pistil with parietal placentation.

Pepper Fam. Solanaceae, *Capsicum annuum.* A vegetable and used as a spice after drying. Black pepper is the dried berry of the pepper wine, *Piper migrum.*

Peppermint Fam. Labiatae. *Mentha piperata.* Is source of peppermint oil and menthol.

Peptide bond A chemical bond (CONH) linking amino acid residues together in a protein.

Perennation For the purpose of living for many years, the stem of certain plants develop underground stems and lodge their permanently.

Perennial irrigation Irrigation methods that provides water for cultivation throughout the year, such as projects based on dams and underground resources that are not subject to seasonal fluctuations.

Perianth When the calyx and the corolla are more or less similar in shape and colour; not clearly distinguishable, they are together known as perianth.

Periblem This lies internal to the dermatogen and is the middle region of the spical meristem.

Pericarp A fruit consists of two portions *viz.* the pericarp developed from the wall of the ovary and the seed developed from the ovules.

Pericentric inversion If the inversion includes the centromere, it is called pericentric inversion.

Perichaetial leaves The involucral leaves surrounding the archegonia (Bryophyta) are these leaves.

Perichaetium In *Marchantia,* surrounding a group of archegonia, a curtain like outgrowth is found. This is the involucre and is known as pseudo-perianth or perigynium.

Periclinal chimera A plant made up of two genetically different tissues, one being external to, and surrounding the other. Also *See* graft chimera.

Pericycle It lies between the endoderm is and the vascular bundles and is composed of several layer of cells. It is partly parenchymatous and partly sclerenchymatous.

Periderm It is the secondary ground tissue.

Peridium In *Aspergillus,* soon after the plasmogamy and the formation of asci the stalk cells of both the sex organs branch freely to produce a large number of sterile hyphae, which coil themselves around the archicarp to form a distinct sheath called peridium.

Perigonal leaves In Bryophyta, the involucoral leaves of a male branch.

Perigynium This is a cup shaped outgrowth formed as a protective covering of the involucre and is known as pseudo-parianth or perigynium.

Perigyny In some flowers, thalamus grows towards around the ovary in the form of a cup carrying on its rim the sepals, petals and stamens. Such flowers are perigynous, and the ovary in them is half-inferiors.

Perinnial The aerial parts of such plants die down every year after flowering or in winter, and a fresh life begins after a few showers of rain when the underground stem puts forth new leaves.

Periplasm During the development of oogonium in *Albugo candida,* the inflated hyphal tip, destined to form oogonium, form a cross wall below the inflation. In the beginning, the cytoplasm is evenly vacuolated and the nuclei are uniformly distributed. Soon it differentiates into an outer periplasm and the inner denser ooplasm (oosphere).

Perisperm The nucellus persists and grows into a nutritive tissue like the endosperm, and is called the perisperm.

Peristome In Bryophyta, in between the operculum and body are present peristome and annulus.

Peritrichous When a larger number of flagella cover the whole body of the bacterium.

Permanent wilting percentage This is percentage of water content of the soil at the time when plants growing in it first show permanent wilting. It is the water content of the soil at the time when plants growing in it first shows permanent wilting. It is different from temporary wilting.

Personate In the personate corolla, the lips are placed so near to each other as to close the mouth of the lower lip.

Petal *See* corolla.

Petiole Stalk of the leaf.

Petite mutant An extremely slow growing mutant of yeast in which oxygen requiring respiration is defective.

***Petunia* sp.** Fam. Solanaceae. Ornamental.

Phages Bacterial virus, which infect, grow on and kill bacterial germs are known as bacteriophages.

Phagocytosis A process by which the cell engulfs solid food particles.

Phanerogams According to the natural system, the plant kingdom is divided into two divisions *viz.* cryptogams or flowerless plants and phanerogams or flowering plants. One of the sub-kingdoms of the plant kingdom.

Pharmacogenetics Biochemical genetics associated with genetically controlled variations in response to drugs.

***Phaseolus* sp.** Fam. Papilionaceae. Have many useful pulses in the genes.

Phaseolus vulgaris Johannsen by working on the pure line theory improved this crop plant.

Phellem Like the vascular cambium, cork cambium cuts off cells both towards and inside and outside. Those formed towards outside mature into cork or phellem. The cells formed by cork cambium towards outside mature into cork or phellem.

Phelloderm Same as secondary cortex.

Phellogen *See* cork cambium.

Phenotype The external appearance of an organism. Phenotype is the result of the interaction between genotypes and environment.

Phialopore In Algae, the plakae is converted into a hollow ball like colony with a small opening called phialopore.

Phillip *See* ovary transplantation experiment.

Phloem Is another conducting tissue, also called bast. It is composed of sieve tubes, companion cells, phloem parenchyma and bast fibres.

Phloem ray That part of the ray which lies in the xylem is wood ray and remaining part which lies in the phloem is the phloem ray.

***Phlox* sp.** An ornamental herb.

Phoenix sylverstris *See* palms.

Phosphoglyceric acid (PGA). This is the first stable intermediate product formed in photosynthesis.

Phosphorescent glow *See* bioluminescence.

Photolysis of water During reaction of photosynthesis, energy is absorbed by chlorophyll and utilized by it in breaking up water, the process being called photolysis of water.

Photosynthesis It consists in the building up of simple carbohydrates such as sugars in the green leaf by the chloroplast in presence of sunlight. Also *See* mechanism of photosynthesis.

Phototaxis It is the movement of an organism in response to light. All free moving algae are phototactic and they move towards the weak source of light.

Photoreceptive *See* eye spots.

Phototaxix It is the movement of an organism in response to light.

Phototropism *See* heliotropism.

***Phragmites* sp.** Fam. Gramineae. The giant reed. Used for making baskets.

Phragnoplast A complex form due to assemblage of residual polar spindle microtubles between the two daughter nuclei during cytokinesis in plants.

Phylocactus A green flattened cylindrical xerophytic stem or branch of unlimited growth.

Phylogeny The evolutionary development of a species or other taxonomic group.

Phyllotaxy The term is used to show the various modes in which the leaves are arranged on the stem or the branch.

Physallis peruviana Fam. Solanaceae. Gooseberry. Edible fruit.

Physiologic race Pathogens of the same species and variety which are morphologically similar but which differ in their pathological characteristics.

Physiology It is the study of the functions of the various parts of the organism and the different life phenomena going on in the living organism.

Phytochrome This is a pigment that exists in plants, is a photoreceptive pigment capable of existing in two forms–one form of pigment absorbs red light, the other absorbs far-red light.

Piliferous layer *See* epiblama.

Pilu Also known as Salvodora, *Salvodara oleoides*. A common tree in the Indian desert bears some kind of edible fruits.

Pinacocytosis A process by which the cell engulfs droplets or liquids.

Pine Order Coniferales. Fam. Pinaceae. Pinus sp. *Pinus longifolia* and *P. excelsa* (blue pine) are the common species.

Pineappble *See Ananas comosus.*

Pink *Dianthus* sp. Ornamental plant.

Pinnae *See* palmate The pinnately compound circinate leaf of ferns further divides in leaflets called pinnae.

Pinnate *See* palmate.

Pinnatifid When in a pinnate type, the incision of the margin is half way or nearly half way down towards the midrib.

Pinnatipartite When in a pinnate leaf, the incision is more than half way down towards midrib.

Pinnatisect When the incision is carried down to near the midrib.

Piper nigrum Black pepper, is the dried berry of pepper vine.

***Pistia* sp.** Water lettuce. A submerged plant.

Pistil Gyanecium or carpel of a flower. In an angiospermic flower, in the centre enclosed by the androecium are one or more carpels which represent the female sex organs collectively known as Pistil or gyanaecium.

Pistillate The unisexual flower having only pistil.

Pistillode A functionless sterile pistil.

Pisum sativum Mendel's experimental plant, the pea.

Pitcher plant *Nephenthes khasiana.* A carnivorous plant.

Pith Same as medulla. Pith or medulla form the central core of the axis. It is composed of large thin walled parenchymatous cells usually with small or large intercellular spaces.

Pithecolobium saman Fam. Mimoseae. A shade tree.

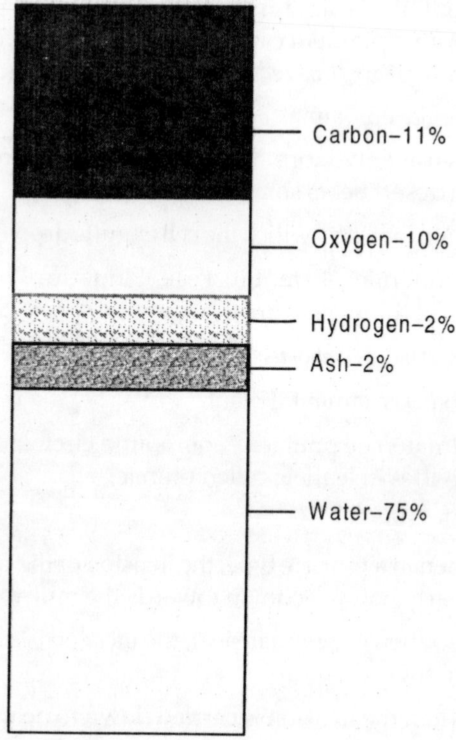

A General Composition of Green Plant Tissue

Pits The unthickened areas in the inner surface of the cell wall, also called pitted thickening.

Placenta It is a ridge of tissue in the inner wall of the ovary, bearing one or more ovules. The manner of distribution of the placentae within the ovary is called placentation.

Plakae In algae, each gonidium undergoes a long series of longitudinal divisions of form 2,4 and 8 cells, which get themselves arranged in the form of a curved plate called plakae.

Plantago pumilla Fam. Scroplulariacea. A common weed of the fields, known as flea seed.

Plant breeding The prime object of plant breeding is to develop varieties that are not only basically higher yielding because of greater physiological efficiency but also more resistant to diseases, insects, draught etc. An equally important object is also to evolve superior varieties that yield agricultural products of high quality. Plant improvement may be affected by the (1) selection (2) introduction (3) hybridization (4) mutation and (5) polyploidy.

Plant exploration Valuable introductions are possible by organising expeditions not only to the primary centres of origin of cultivated plants but also to unexpected regions of the world.

Plant introduction The plants cultivated by modern man have all descended from wild ancestors. These have to be brought to various locations for cultivation by introduction through the agency of man: The introduced material(s) can be used in three ways (1) They can be directly used by increasing it en masse. (2) Using the desirable strains from the introduced material. (3) By using the introduced material as a parent to the locally adopted varieties.

Plant succession A progressive sequence of changes in the vegetation of an area. In all environments, succession begins with a pioneer stage in which plants colonise a bare surface. These plants modify the environment (by introducing humus) so that slightly more demanding species are able to replace them. Each stage of succession improves the existing plant community, improves the soil and other conditions so that the next more demanding plant community can succeed it.

Plasma genes The units of heredity located in cytoplasm.

Plasma membrane The semipermeable membrane enclosing the cell. Also called cell membrane.

Plasmodesmata The minute pores (thin areas) often occur in the cell wall known as plasmodesmata.

Plasmogamy Fusion between the protoplasts of gametes without nuclear fusion.

Plasmogenes Also called cytogenes. A vast majority of heritable variations are nuclear but it has been found that cytoplasm contains extranuclear units of inheritance called plasmogenes.

Plasmolemma The protoplasm is surrounded by differentially permeable membrane called plasmalemma.

Plasmolysis Loss of water from the cell contents results in the concentration of the protoplasm away from the cell wall. The cell is now said to be plasmolysed and the process by which this state is achieved is plasmolysis.

Plasmon The sum of the genetic properties of the cytoplasm in a species.

Plastids Plant cells contain plastids of various kinds, the most important of which are the chloroplasts, which are the centres of photosynthesis activity. These are protoplasmic bodies of various shapes present in the cytoplasm.

Pleiotropic gene When a gene causes changes in two or more parts of characters that are not obviously related, the gene is called pleiotropic.

Pleiotropism The control of more than one character by a single gene.

Plerome This lies internal to the periblem and is the central region of the stem apex.

Ploidy Variation of number in whole chromosome.

Plumeria rubra Pagoda or temple tree.

Plumule The part of the axis lying towards the pointed end of the seed is called radicle, while the other end lying in between the two cotyledons is known as the plumule.

Pneumatophore A special type of roots. These are respiratory roots which develop from underground root.

Pod Same as legume.

***Pogostemon* sp.** Fam. Labiateae. Patchouli, yields Patcholi oil.

Poinsettia Fam. Euphorbiaceae, *Euphorbia SP.* The bracts are brightly coloured and attractive.

Point mutation A heritable change taking place at a single gene locus.

Polar bodies The expelled products of the division of oocyte nucleus in animals during oogenesis.

Polar nuclei in Angiosperms, the embryo sac gets one nucleus from each pole moves towards the centre and is known as the polar nucleus.

Pollen chamber The nucellar beak in Gymnosperms has a narrow cavity called the pollen chamber.

Pollen drop In Gymnosperms, the germination of the microspore after dispersal are caught by the pollen drop (A small droplet of mucilagenous liquid exuded by the micropyle).

Pollen grain The anther bears four chambers or pollen sacs, each filled with a granular mass of small (male) spores called grains.

Pollen mitosis The nucleus of the pollen grain divides by mitosis (first pollen mitosis) into two, one called the tube nucleus and the other called the generative nucleus.

Pollen mother cell In each chamber or locule of the anther called pollen sac, several cells in the centre enlarge and become microsporocytes or microspore mother cells or pollen mother cell (P.M.C.).

Pollen tube When the pollen germinates the intine grows out into a tube, called pollen tube.

Pollination It is the transference of pollen grains from the anther of the flower to the stigma of the same flower or the another flower of the same or sometimes allied species. The act of transfer of pollens from the anthers to the stigma.

Pollination contrivances By cross pollination, better seeds and healthier offspring are normally produced. Nature therefore favours this process and helps it by certain contrivances in flowers, which wholly or sometimes partially prevent self-pollination. It must however be noted that in many flowers, there is still provision for self-pollination if the other method fails.

Pollinator In hybrid production of cross pollinated crops, the male sterile line is used as female line and the male fertile line is used as the pollinator (Maize).

Polliniuin In some plants (Orchids), the pollen cells are not free but become united in a mass known as pollinium.

Polyarch When the metaxylem elements of different bundles meet in the centre and the xylem bundles give a star shaped appearance.

Polyadelphous stamens When the filaments are united into a number of bundles (more than two), but the anthers are free.

***Polyalthia* sp.** The mast tree. Ornamental.

Polyanthes tuberosa Fam. Amaryllidaceae, tuberose.

***Polycarpon* sp.** Has a curvex ovule.

Polycyclic In Gymnosperms, as many as 14 consecutive rings of secondary xylem and phloem can be seen formed by successive cambial rings. Hence the secondary wood in *Cycas* has been described as polycyclic.

Polyembryony As there are a few archegonia, (pines) in the ovule some more embryo may be formed. This is called polyembryony. This is a source of halploids in some cases. This condition is sometimes due to the occurrence of more than one embryo sac within the ovule. Cause variation in the seedling plant as found in Citrus and mango.

Polygenes Minor genes controlling quantitative characters which individually have too small effect to show clear segregation. Also *See* multiple factors.

Polygenic inheritance This assumes that there is a series of independent genes for a given quantitative trait. Dominance is usually incomplete but these genes are cumulative or additive in their effect. This type of inheritance governs the polygenic characters or the quantitative characters.

Poly-haploids These are haploids which arise from polyploids *e.g.* haploid of *Triticum vulgare* with one representative of each chromosome of the A, Band D genomes.

Polyhybrid It is hybrid resulting from a cross between parents differing in several genes.

Polymer A chemical compound composed of two or more units of the same compound.

Polymeric genes Same as multiple genes or poly-genes.

Polymorphic locus A locus at which the most common allele has a frequency of less than 0.95.

Polypeptide A compound containing the amino acids residues joined by peptide bonds. A protein may consists of one or more specific polypeptide chains.

Polypetalous A condition when the petals are free.

Polyphllous When the whorl are free.

Polyploid These are evolved by the involvement or changes in the number of entire sets of chromosomes.

Polysepalous When the sepals are free.

Polysomaty A form of endopolyploidy in which chromosomes separate from each other within the nucleus, each chromosome is normal in appearance and the total polyploid number of chromosome is distinguishable.

Pome This is an inferior, two or more celled, fleshy syncarpous fruit surrounded by the thalamus.

Pomegrtanate Fam. Punicaeae, *Punica granaturn.* A fruit plant.

Poppy *Papaver* sp. Fam. Papaveraceae. Latex vessels are found in this plant *e.g.* Opium poppy, garden poppy and prickly poppy and also in some species of sunflower family *e.g. Sonchus* sp.

Population method Same as bulk method of breeding.

Portia tree Fam. Malvaceae. *Thespetia sp.* Shade tree.

***Portulaca* sp.** Used as a leafy vegetable. Has a medicinal value. A serious weed of cultivated fields.

Potato Fam. Solanaceae. *Solanum tuberosum.* The tubers are used as vegetable worldwide.

Potentilla fulgens Fam. Rosaceae. The silver weed.

Potometer (Darwin's) This is U-tube potometer for measuring transpiration rate in plants. Other potometer used for this purpose is the Ganong's potometer.

Preprophase band A narrow bundle of microtubules situated just beneath the plasma membrane before the onset of mitosis.

Primary centre of origin In the Andes in South America, there existed all ranges of variations in potato crop-from a perfectly tuber forming cultivated type of potato to the non-tuber forming wild types. Such regions are called primary centre of origin for crop. On this basis, it was said that all the cultivated crops of the world originated in such centres and later spread to different parts of the world.

Primary constriction The chromosome has a non-staining constriction called the centromere or primary constriction.

Primer A short nucleic acid segment, hydrogen bounded to a parental DNA strand at which replication of DNA is initiated.

Primordial utricle In Algae, a large nucleus is embedded in the cytoplasmic layer enclosing a central vacuole is called primordial utricle.

Primrose It is an example of dimorphic heterostyly.

Primula In this plant, there are two types of flowers borne on different plants. It is an example of distyly.

Principle of limiting factor When a process is conditioned to its rapidity by a number of separate factors, the rate of the process is limited by the pace of the slowest factor.

Procambium These groups are strands of elongated cells which form the procambium. This is situated at a little distance behind the apex of stem.

Proembryo After the quarter stage-nuclei, they move to the bottom of the egg cell and divide again into 8 nuclei. Walls now appear between them. Further division result in the formulation of four tiers of four cells each; this 16-celled stage is called the proembryo.

Progeny test Evaluating the genotype of an individual based on the performance of its offspring.

Promeristem The apical meristem of growing region is composed of a mass of small usually rounded or polygonal cells which are essentially alike and are in a stage of division. These constitute the promeristem.

Promoter site Site on DNA at which RNA synthesis is initiated in an operon. May be combined with a repressor to inhibit RNA synthesis by associated cistrons.

Prop or still root A number of roots are produced from the main stem and often from the branches. These roots grow vertically or obliquely downwards and penetrate into the soil.

Prophase The state in mitosis and meiosis characterised by alternations in the nucleus before the nuclear membrane is dissolved.

Prosopis spedigera Fam. Mimoseae. A waste land plant.

Protandry When the anthers mature earlier than the stigma of the same flower.

Protenor A squash bug, the females have 14 chromosomes and the male has only 13. A condition of XX-Xo type.

Proteins These are the most important compounds found in the plants. Different proteins have different chemical formula *e.g.* Zein from Indian corn has the formula C_{736} H_{1161} N_{181} O_{208} S_3; gulletin from wheat is C_{685} H_{1068} N_{196} O_{211} S_5 etc.

Prothallial cell In Gymnosperms, the first division of the microspore during the development of the male gametophyte results into the formation of a small prothallial cell which does not disorganise and a large antheridial cell.

Prothallus The gametophyte in fern, a very small green flat heart-shaped body. Under favourable conditions of moisture and temperature, the spore in Pteridophytes, the spore germinates, the outer thick exposure ruptures and the inner thin endospore grows out to form a short papilla. This papilla then grows, divides repeatedly in different plants ultimately gives rise to a heart shaped structure–the prothallus.

Protista Organisms include under protista in modern classifications posses nuclear membranes around their nuclei. This type of organisms can be both unicellular as well as multicellular.

Protococcus Is a common unicelluar alga, terrestrial in habit.

Protoderm When the apical meristem is not sharply distinguished in the three parts, the protoderm corresponds to the dermatogen.

Protogyny When the pistil matures first than the anthers in the same flower.

Protonema In moss, the spore grows into a green much branched filament known as protonema.

Protophloem The outer portion of phloem consisting of narrow sieve-tubes is the first formed phloem or protophloem.

Potoplasm The exact chemical composition of the protoplasm cannot be determined for various reasons. However, the analysis of dead protoplasm reveals a long list of elements present in it.

Protoxylem The first formed xylem.

Prune A rosaceous fruit. The genus *Prunus.*

Protons An atom is made up of two parts (1) an inner positively charged nucleus consisting of positively charged particles, the protons and neutral particles called neutrons (2) An outer constellation of negatively charged electrons. The charges are so balanced that an atom is electrically neutral.

Protoplast A structural unit of protoplasm; all the living (protoplasmic) material of a cell. The two principal parts are nucleus and cytoplasm.

Prunus This fruit plant is faced with gametophytic incompatibility sometimes in some varieties.

Pteris A fern plant.

Pteridophyta The plant body is differentiated into the stem, leaves and roots; there is regular alternation of generation. They are called vascular crytogams and include ferns.

***Pterospermum* sp.** Ornamental plant.

Ptomaine This is a kind of poisoning caused by the bacteria *Clostridium botulisum* especially food poisoning.

Pulsation In plants, the spontaneous movement of variation is very remarkably exhibited by the pusation *i.e.* rising and falling the two later leaflets of the Indian telegraph plant *(Desmodium gyrans).*

Pulses Such food plants which are used as vegetable protein in human diet, their roots have nodules inhabited by bacteria for fixing atmospheric nitrogen.

Putrification When proteins are decomposed certain bad smelling substances like Methane hydrogen sulphide and organic acids are produced and the process is called putrification.

Pyrenoids Associated with the chloroplasts are a large number of small starch bodies called pyrenoids.

Pyrus Same as for Prunus.

Pulvinus The swollen leaf base.

Pummelo Also called shaddock. Fam. Rutaceae. *Citrus sp.*

Pumpkin Fam. Cucurbitaceae. *Cucurbita pepo.* Vegetable plant.

***Pupalia* sp.** Bears hooked bristles borne by the perianth.

Pure line selection A large number of individual plants are selected from the genetically variable original population. (2) From each of the selected individual plants, a 'progeny row' is raised and unsuitable types are discarded on the basis of visual observations. (3) The remaining selections are compared for their yielding ability and other aspects of performance not only with each other but also with 'standard' varieties in randomised replicated yield trials for at least three years.

Pure line theory A strain in which all members have descended by self fertilization from a single homozygous individual.

Purine Nitrogen compounds as occurring in DNA and RNA; these are adenine and guanine.

Pyrimidine Nitrogenous base occurring in DNA (thymine and cytocine) or DNA (uracil and cytosine).

Pyrus malus The edible fruit. Fam. Rosaceae, apple.

Pyruvic acid Is a product of the reaction during anaerobic and aerobic respiration.

Quadrivalent A group of four homologous chromosomes held together by chiasmata during the period from diplonema to first metaphase.

Qualitative character A character in which variation is discontinuous.

Quantitative character A character in which variation is continuous making separation into distinct classes difficult.

Quantitative inheritance The individual of each homozygous parental line have the same genotype and therefore the two lines to which the parents belong would show very little variability within themselves. The phenotypic differences between individuals within a parental line are only due to environment.

Quantum Although light energy travels as an electromagnetic radiation embodying wave properties, such as wave length and frequency, it is nevertheless emitted and absorbed in the form discrete packets known as quanta.

Quiscent centre An inert place between root tip cap cells where there is little or no division in the meristematic cells or there is a restricted division only.

Raceme An inflorescence in which the main axis is elongated and it bears laterally a number of flowers which are all stalked, the lower or older flowers having longer stalks.

Racemose In this type of branching, the growth of the main stem is indefinite *i.e.* it continues to grow indefinitely by the terminal bud and give off branches laterally in acropetal succession.

Rachis The midrib of a pinnately compound leaf. Also the spikelet bearing stalk in the Gramineae. The cross section of rachis of *Cycas* shows that it is almost biconvex in outline. In Gramineae, the flowers are arranged firmly on the rachis.

Radicle The root primordia. It is surrounded by a sheath (coleorhiza) at one end and tip of plumule at the other end.

Radiomutation A mutation resulting from exposure of living tissue chromosomes to particulate radiation such as X-rays, gamma-rays and fast neutrons.

Radish Fam. Cruciferae. *Raphanus sativus.* A vegetable plant of which roots and leaves are used.

***Rafflesia* sp.** A total root parasite found in East Indies, bears the biggest flower in the world.

Railway creeper Fam. Convolvulaceae. *Ipomea palmata.* Waste land plant of ornamental use.

Ramenta In ferns, the numerous brown scales covering the stem and petiole.

Randomized block design An experimental design in which the treatments are arranged in random order within the blocks or replicates.

Rangoon creeper *Quiscalis* sp. Gives scented flowers. The arrangement of the leaves is said to be superimposed.

Ranunculaceae A family of angiosperms.

***Ranunculus* sp.** Fam. Ranunculaceae. Butter cup plant.

Rape Fam. Cruciferae. *Brassica napus, B. juncea.*

Raphe In the ovule, the food is carried to the nucellus, by raphe.

Raphides These are needle like crystals occurring singly or in bundles. They are in most of the plants.

Raspberry Fam. Rosaceae. *Rubus ideaeus.* Wild raspberry is *R. molluccanus*. An edible fruit.

Rattlewort Fam. Papilionaceae. *Crotalaria sericea.* Good for use, medicinal.

***Ravenala* sp.** Traveller's tree.

Readthrough Transcription beyond a termination sequence due to failure of ribonucleic acid polymerase termination factor to recognise the termination codon.

Recapitulation theory Haeckal's biogenetic theory that an organism passes through developmental stages resembling various stages in the phylogeny of its group. Ontogeny repeats phylogeny.

Receptacle The circular disc bearing some special erect male reproductive branches, antheridiosphores, in *Marchantia.*

Receptor site In the first phase, the tip of the virus tail becomes attached to the cell wall or surface membrane at a specific place known as receptor site.

Recessive allele That form of a gene which does not express itself in the presence of the contrasting (dominant) allele. Mendel recognised the presence of constant differentiating characters.

These contrasting characters are attributed to the presence of allelemorphs (allelon-one another; morphus-form) or allelic genes of alleles. In case of peas, the dominant gene is represented by a capital letter and its recessive allele by the corresponding small letter. Thus the gene responsible for round seed coat is represented as R and the allele responsible for wrinkle is r.

Recessive mutation A recessive change does not appear in the individuals produced by the mutated gametes but will appear in about one quarter of the offspring of those plants, if they are self fertilised.

Reciprocal cross Two crosses between two plants or strains in which the male parent of one cross is the female parent of the second cross *e.g.* A × B and B × A.

Reciprocal translocation Also called 'illegitimate crossing over' involves the mutual exchange of segments between non-homologous chromosomes. If two normal nonhomologous chromosomes have a sequence of regions designated ABCDEF and KLMNOP, reciprocal translocation might rise to ABCNOP and KLMDEF.

Recombination The new association in one individual of phenotypic traits from both parents *i.e.* a non-parental individual. Also random combinations of genes from different linkage group resulting from meiosis.

Recon The smallest segment of DNA or unit of a cistron that is capable of recombination; probably as one nucleotide pair.

Recurrent parent The parent of a hybrid with which it is again crossed or with which it is repeatedly crossed; the back cross parent.

Red wood Fam. Papilionaceae. *Dalbergia sissoo.* Timber tree.

Reductional division In the higher plants, meiosis takes place in the ovaries and anthers. Meiosis comprises of two divisions of the nucleus but only one division of the chromosomes. The first division is called reductional division (or meiosis I) because the reduction in chromosome number occurs during this division and the second division is called equational division (meiosis II) because it brings about the equal separation of sister chromatids into the daughter nuclei.

Reed Fam. Gramineae. *Phragmites* sp. The giant reed is *Arundo* sp.

Regeneration The replacement of tissue and organs by an organism which have been lost or severely injured.

Registered seed The progeny of foundation seed normally grown to produce certified seed.

Regma A dry, three to many chambered fruit developing from syncarpous ovary.

Rejuvenescence Return of the old protoplasm of a cell to a youth condition again.

Relation coiling The chromatids are closely expressed to each other throughout their length and are at the same time twisted around .each other like the strands in a rope. Such a coiling or twisting of pairs of chromatids is known as relational coiling.

Relative humidity It is the amount of water vapour present in the air expressed as a percentage of the total amount of water vapour required to saturate it at a given temperature:

$$\text{R.H.} = \frac{A}{T} \times 100$$

The lower the relative humidity of the air greater will be the rate of transpiration.

Replum A false partition wall that develops later in a siliqua fruit.

Replicate To form replica; applies to synthesis of new DNA from pre-existing DNA as a part of nuclear division.

Replication Replicating circular DNA molecular of viruses, chloroplasts and mitochondria schematically appear like the Greek letter theta.

Replication origin The position along molecule at which DNA replication begins.

Repressor A protein produced by a regulator gene that can combine with a repress action of an associated operator gene.

Reproduction A method by which an organism reproduces itself in order to maintain the continuity of species.

Reps The common dosimery determination of radiation by neutrons is in reps which is Roantgen equivalent physical.

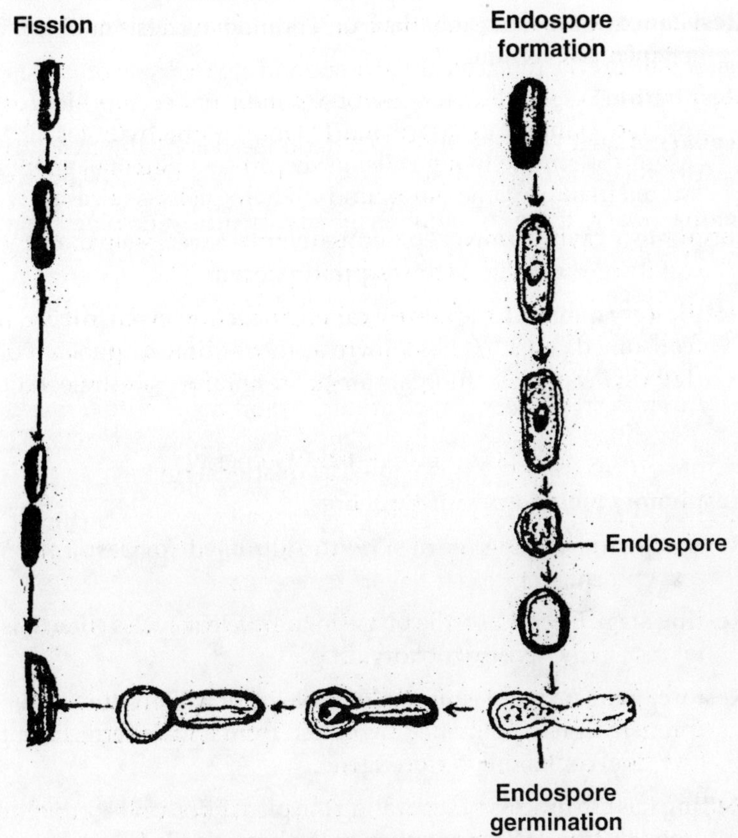

Mode of Reproduction in Bacteria

Repulsion The condition in linked inheritance in which a individual heterozygous for two pairs of linked factors received the dominant member of one pair and the recessive member of the other pair from one parent and the reverse condition from the other parent.

Resins These are chemically complex organic substances found in the stem of conifers and occur in abundance in special ducts, the resin ducts.

Resistance Method of breeding varieties resistant to diseases do not differ greatly from those used for other characters. If genes for resistance occur in commercial varieties, selection within these varieties will lead to the evolution of resistant strains.

Resistance factor A plasmid that confers antibiotic resistance to the recipient bacterium.

Respiration Essentially a process of oxidation or decomposition of organic compounds, particularly simple carbohydrates such as glucose in the living cells. In the process potential energy stored in the organic compounds in living cells is released.

Respiratory cavity In lower epidermis internal to each stoma, a large cavity is found; this is the respiratory cavity.

Respiratory quotient The ratio of carbon dioxide evolved to oxygen consumed *i.e.* CO_2/O_2 is known as the respiratory quotient or R.Q. of the process. If sugars are the respiratory substrate, R.Q. = 1

$$C_6H_{12}O_6 + 6CO_2 + 6H_2O + \text{energy}$$

Respiratory roots *See* pneumatophores.

Respiroscope A flask with a bent bulb used for respiration experiment.

Resting stage Interphase; the phase in the nuclear cycles when it is at rest, in the sense of not dividing.

Restorer gene A cytoplasmic male sterile plant will produce fertile progeny if it is pollinated by pollen from a male fertile plant ('R' line) containing restorer gene.

Retting It is the process of separating the plant fibres and is brought about by the combined action of the bacteria *(Bacillus subtilus)* and the species of *Mucor.* They secrete pectinose enzymes which dissolve the middle pective lamallae of the cell walls and the fibres are set free.

Reverse mutation The mutation of a mutant gene back to its original state.

Rhea or ramie *Bochmeria* sp. A fibre bearing plant.

Rhizoids in Riccia, the thallus is small and flat with a longitudinal groove on the upper surface along the midrib, and a number of slender unicellular hair-like structures are rhizoids.

Rhizobium radicicola These bacteria have the power of fixing the free nitrogen of the soil air in the nodules.

Rhizome It is a thickened prostrate underground stem.

Rhizophora Typical mangrove plants come in this category.

Rhizophore A Selaginella, a long slender root-like organ is given off from the stem at the point of bifurcation. This is the rhizophore, the root bearer.

Ribonucleic acid (RNA) A single stranded nuclei acid molecule, synthesised principally in the nucleus from deoxyribonucleic acid composed of a phosphate-ribose backbone with purines (adenine and guanine) and pyramidines (uracil and cystocine) attached to the sugar ribose. RNA is of several kinds and functions to carry the genetic message from the nuclear DNA to the ribosomes.

Ribonucleotide Portion of RNA molecule composed of one ribose-phosphate unit plus a purine or a pyrimidine.

Ribose nuclei acid *See* ribo nucleic acid and RNA.

Ribosome A cytoplasmic structure usually adherent to the endoplasmic reticulum, which is the site of protein synthesis.

Riccia It is a rosette type of thalloid liverwort showing distinct dichotomous branching.

Rice *Oryza sativa*. Fam. Gramineae. Paddy. A staple food grain of the majority of people world over.

Ringed bark *See* scaly bark.

RNA Analysis of RNA (Ribonucleic acid) content of the cell shows that it is found mainly in the nucleolus, the ribosomes, the cytoplasmic (or endoplasmic) reticulum and free in the cytoplasm. RNA is regarded as an intermediary between DNA and the metabolism of the cytoplasm.

DNA —————— RNA —————— Protein

Transcription Translation

RNA messenger Or template RNA. The DNA template on the chromosome is the blue print for the synthesis of the messenger or the template RNA, also known as ribosomal RNA.

RNA transfer The cytoplasmic RNA, so called transfer RNA, consists of only of a smaller number of nucleotides (about 80) and a possibly synthesised from other RNA. It plays as important role in protein synthesis as it assists in the conveyance of amino acids to the correct site on the messenger or template RNA.

RNA nucleolar The nucleolus has always been a puzzling thing for cytoplasmic studies. Its disappearance during cell division and reappearance during metaphase suggest that it might be some sort of nucleic acid storage system but even now there is no clear evidence as to function of this organelle.

Rod shaped chromosome *See* J-shaped chromosome.

Roentgen unit X-rays electro-magnetic radiations with a wave length of 0.06 angstron. The amount of X-ray treatment is measured in Roentgen units (r units).

Rogue The undesirable individuals from a variety in the field.

Rolling alga The locomotion is brought about by the combined action of the cillia of all the cells. The colony shows a characteristic rolling movement and is therefore called rolling alga.

Rolling circle replication A process in which replication proceeds around the circle from point of origin by adding nucleotides to the 3′ end with simultaneous rolling out of the 5′ end as a tail of increasing length.

Root It is the descending organ of the plant and is originally the direct prolongation of the radical of the embryo.

Root cap The root is covered at its tip by a kind of cap called root cap.

Root crop Crop of such plants which are cultivated for production of roots, later to be used as vegetable medicines or industry. Such crops are radish, turnip, beet and carrot etc.

Root hair Externally at its basal portion, this region produces a cluster of very fine and delicate thread-like structures known as root-hairs.

Root pocket In water plants, a loose sheath which comes off easily is distinctly seen at the apex of each root. This is an anomalous root-cap, called root pocket.

Root pressure The water with mineral salts absorbed from the soil by the root hairs gradually accumulates in the cortex. An intermittent pumping action naturally gives rise to a considerable pressure. Thus the water is forced into the xylem and to the upper region of plants.

Root primordium In *Cycas*, an embryo develops into a short axis with a radical or root primordia.

Rosaceae A family of angiosperms.

Rosaceous This form of corolla consists of five petals, with very short claws or none at all, and the limbs spread regularly outwards.

Rose Fam. Rosaceae. *Rosa* spp. Dog rose is *Rosa canina*; wild rose is *R. gigantea;* Bengal rose is *R. damascena* and *R. centifolia* and musk rose *R. moschata.*

Rosemary *Rosamarinus* sp. Yields oil of rosemary.

Rotate When the tube of the corolla is comparatively short and its limb is at right angle to it, the corolla having more or less the appearance of a wheel.

Rozelle Fam. Malvaceae. *Hibiscus subdariffa.*

Rubber This is obtained for the latex of *Hevea brasiliensis.* The latex is collected by tapping the bark.

Rubiaceae A family of angiosperms.

Rubia cordifolia Fam. Rubiaceae. A plant used in dye industry.

Rubus ideaus Fam. Rosaceae. Raspberry.

***Ruellia* sp.** Has fasciculated roots.

***Rumex* sp.** The sorrel. Has orthotropous ovules.

Runner A sub-aerial modification of stem. This is a slender, prostrate branch with long internodes creeping on the ground and rooting at the nodes.

***Ruscus* sp.** Fam. Liliaceae. Butcher's broom.

Rush *Juncus sp.* scorpoid cyme of branching.

Rutaceae A family of angiosperms. Includes all Citrus fruit plants and also others of economic importance.

Saccharum officinarum Fam. Gramineae. Sugarcane plant. Yields cane sugar.

Saccharum spontaneum This plant has been used in the improvement of sugarcane by breeding methods.

Safflower Fam. Compositeae. *Carthhamus tinctorius.* An important oil yielding plant.

Saffron *Crocus sp.* Medicinal plant.

Sagittate When the blade is shaped like an arrow.

Salmonella *See Escherichia coli.*

Samara This is dry indehiscent, one or two seeded, winged fruit developing from a superior ovary.

Sandalwood *Santalum album.* The heartwood yields 5 to 7 per cent of a yellow aromatic volatile oil.

Sap wood *See* Albumum.

Saprophytes These are free living forms and include moulds, bacteria and fungi. Any non-living organic component provides usable food for these.

Saraca indica Fam. Caesalpinieae. Ornamental plant.

Sarcina Cubes or packets of cells. In the case of division, these bacteria are in all planes but in a regular order.

Sat-chromosome A chromosome possessing one or more satellites.

S_0 S_0 symbol used to designated the original selfed plants.

S_1 S_1 S_2 symbols used for designating first selfed generation (progeny of S_0 plant), second selfed generation (progeny of S_1) etc. The second generation is called S_2 and so on.

Satellite A short segment of chromosome separated from the main dry by one or more constrictions.

Saturation deficit The rate of transpiration depends upon the difference between the amount of water vapour actually present in the air and the amount necessary to completely saturate it–the saturation deficit.

Scalariform This is a form of lignin formation in plants. This type is found where the thickening material takes the form of spiral end. Also called ladder-like.

Scale leaves in Gymnosperms, these are small, thin and brown in colour. These are borne on both long and dwarf shoots and are mainly protective in function.

Scaly bark These are two types of barks–ringed bark and scaly bark. If successive cork cambia are in the form of complete rings, the bark falls off as complete circular sheets and is known as ringed bark. And if the successive cork cambia occur as separate strips connected at their ends, the bark falls as small scaly pieces and is known as scaly bark.

Scape When the arial shoot bears either a single flower or a cluster of flower on the top, such a shoot is called scape.

Schizogamous By the enlargement of small intercellular spaces between the cells without affecting the cells themselves in any way. Such cavities are termed as schizogamous.

Schizogenous cavity These are intercellular cavities and form an intercommunicating system.

Schizomycophyta By way of a classification, the bacteria are included in this division.

Schizosaccharomyces *See* Plant Pathology.

Schleiden and Schwann These two German scientists as far back as 1835 enunciated the idea that cells arise only from the pre-existing cell.

Sclerenchyma These cells consist of very long, narrow, thick walled and lignified cells usually pointed at both ends. It is the chief mechanical tissue of plants. The cells are usually long, narrow, pointed at both ends and uniformly thickened by the deposition of lignin.

Sclerotic cells These are special type of sclerenchyma found in the plant body, also called stone-cells.

Scorpiod cyme Alternate sided cyme.

Scutellum *See* cotyledon.

Sears and Mcfadden According to these scientists, the probable origin of the hexaploid wheat may be summarised as:

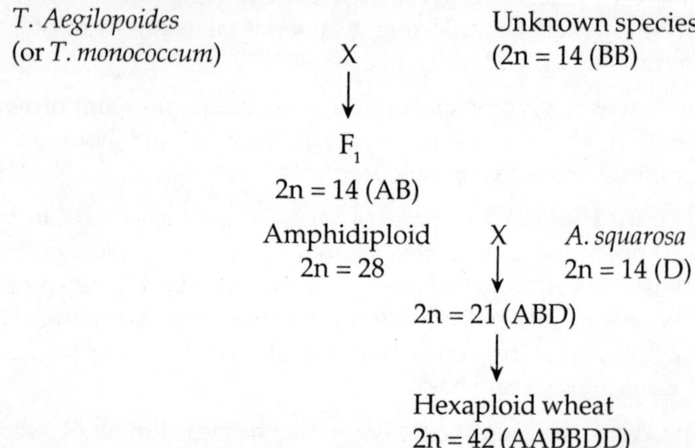

Secale cereal Tricale is an amphidiploid (2n=56) synthesised by Muntzing from a hybrid between wheat *(Triticum vulgare)* 2n=42 and Rye *(Secale cereale)* 2n=14.

Secondary growth It takes place as a result of the formation of new (secondary) tissues in them. The increase in thickness due to the addition of secondary tissue cut off by the cambium in the stelar and extra-stelar regions respectively is spoken of as secondary growth.

Secondary nucleus in Angiosperms, in the embryo sac, the two polar nuclei ultimately fuse giving rise to a diploid nucleus known as secondary nucleus.

Second pollen mitosis As the pollen tube approaches the ovule, the generative uncleus by mitosis (the second pollen mitosis) to form two sperms nuclei or male gametes.

Secretary tissue *See* laticiferous tissue.

Sectorial chimera *See* graft chimaera.

Seed certification The purpose of seed certification is to maintain and make available to the farmers high quality seeds of superior strains of crops, as grown and distributed as to ensure genetic identify.

Seed generation (S.G.) S.G. 1 is used directly produced by cross-pollination from F_1 plants will grow. S.G. 2 represents self-bred seed from an F_1. S.G. 3 is self-bred seed from an F_2 etc.

Segmental interchange An exchange of segments between two non-homologous chromosomes.

Segmental polyploids In contrast to the true polyploids.

Segregation Separation of paternal from material chromosomes at meiosis and consequent separation of genes leading to the possibility of recombination in the offspring.

Seismonasty The movement included by mechanical stimuli. Plants having this tendency develop special structures on their organs.

Selection The choosing of individuals with any specified desired characters from a population with the object of improving the average type.

Selection of parents for hybridization A hybridization programme can be started only after the potential materials. Great care is taken in selecting parents having the desired characters.

Self bred Arising from an ovule, or ovules fertilised by pollen from the same plant.

Self fertilization Functioning of an individual both as male and female parent plants are selfed if sperm and egg are supplied by the same individual.

Self pollination This can be affected by collecting pollen from the terminal tassel and by applying it to the silk of the axillary cobs in the same plant, taking precautions to prevent contamination by foreign or undesired pollen in case of maize. An inbred line can be maintained either by self pollination of by sib-pollination *i.e.* mating the plants within the same inbred line.

Sensitivity Sensitivity in plants is not noticeable as much as in animals. Some plants do show changes in the plant organs in response to certain stimuli.

Sepals The outermost whorl is of small green (rarely coloured) leaf like structure called sepal. Also *See* calyx.

Serology The study of interactions between antigens and antibodies.

Sesbania grandiflora Fam. Papilionaceae. A good green manure plant.

Seta The stalk of sporangium of *Marchantial.* In Bryophyta, the sporogonium at maturity is about 2 to 4 cms long and is divisible into three main regions–Foot, Seta and Capsule.

Set of chromosomes The complement of chromosomes occurring in a gamete.

Sex chromosome A chromosome which has not truly similar homologue in the heterozygous sex and which is closely bound up with sex determination. These do not occur in individual pairs in both sexes in diploid organisms; in man and fruit fly (Drosophilla), these are designated as X and Y chromosomes.

Sex determination A large number of species of animals and a small number of species of plants, egg and sperms are produced by different individuals *viz.* females and males respectively. In most of these dioecious organisms, females and males visibly differ in their chromosomal constitution. The precise form of their chromosomal difference between the sexes is not the same in different organisms. For types–XX-XO; XX-XY; ZO-ZZ; and ZW-ZZ are known.

Sex determiner In the Protenor bug, the odd chromosome of the male determines the sex of the individual receiving it and it was consequently called the sex determiner or the sex-chromosome or the 'X' chromosome.

Sex heterogametic A sperm or egg that containing both types of sex chromosomes such as XV, ZO, XY or XO types of chromosomes.

Sex homogametic A sperm or egg that may contain AA, ZZ types of sex chromosomes.

Sex influenced characters These are the characters which may be expressed differently in the two sexes even when their genotypes are identical. The mere influence of the sex is sufficient to alter the phenotypic expression of a gene in an individual.

Sex influenced genes Genes whose dominance is dependent on sex so that in one sex the heterozygote shows the character concerned while in the other sex the character is recessive.

Sex limited character *See Colias* sp.

Sex linkage The association which exists between sex and genes borne on the sex chromosomes. More than 140 genes in *Drosophila melanogaster* are situated on the sex chromosome. The pattern of inheritance by such genes called sex-linked.

Sex linked character Hereditary character whose genes are located on the sex chromosomes.

Sexual reproduction Reproduction involving union of gametes.

Sheath In some plants, the leaf base expands into a sheath.

Shrubs These are medium inbred line.

Sib Plants of the same inbred sibs.

Sib mating The mating of sibs.

Sibs Progeny of the same parents derived from different gametes.

Sieve plate Each transverse wall is perforated by a number of pores. If then looks very much like a sieve and is called sieve-plate.

Sieve tubes These are slender, tube-like structures composed of elongated cells, placed end on end.

Silviculture The cultivation of trees including the theoretical and practical aspects of forestry.

Single cross A cross between two inbred lines.

Single-cross seed In an isolated plot, the male sterile line A and the restorer line R are grown. For every six rows of A line, two rows of R line are grown.

Smooth rhizoids In Bryophyta *(Riccia)*, all the rhizoids are unicellular and unbranched. In some of the rhizoids, inner surface of the wall is smooth and such rhizoids are known as smooth rhizoids.

Soft bast *See* secondary phloem.

Sol A colloidal solution.

Solanaceae A family of angiosperms. Consists of many economic plants. This is a small family consisting of about 85 genera and about 2000 species which are widely distributed through tropical and temperate regions of the world. Floral formula is:

$$\oplus ⚥ K_{(5)}\ C_{(5)}\ A_3\ \underline{G}_{(2)}$$

Solanum tuberosum Fam. Solanaceae. Potato. A common vegetable tuber.

Solarization It has been demonstrated that intense light has inhibitory action on photosynthesis which is known as solarization.

Soma According to Weismann (1834-1914), the germplasm is handled down from parent to offspring and it gives rise to the body of soma (somatoplasm).

Somatic Pertaining to body. Having two sets of chromosomes, one set normally coming from the female parent and the other from the male.

Somatic cell In the somatic cells (or the vegetative cells) of most organisms that reproduce sexually, chromosomes are found in pairs, of which one member has come from the mother and the other from the father.

Somatic mutation A mutation in a somatic cell, resulting in a chimaeral individual with respect to the mutant character. If the mutation occurs in later stages of somatic development, only those cells derived by divisions from the mutant cell would have the mutant gene.

Somatic number Number of chromosome found in the body cells of an individual (2n).

Somatoplasm Same as soma.

Sorosis A multiple fruit developing from a spike or spadix.

Sorus In fern, on the under surface of the leaf or the sporophyll, a number of dark brown structures, pale green when young may be seen. These are called sori (sing. sorus).

Sources of resistance Breeding programmes for evolving superior variations resistant to diseases can be planned only if genes for resistance have been found in some variety of that crop or some species closely related to it.

Spadix It is spike with a fleshy axis which is enclosed by one or more large, often brightly coloured bracts, called spathes.

Spatulate When the shape of the leaf blade is like a spatula.

Species This is a group of individuals (plants or animals), of one and the same kind.

Sperm The male gamete.

Sperm nucleus Fertilisation consists essentially of the fusion of the embryo sac and sperm nucleus of the pollen grain resulting in the formation of zygote, the first cell of the individual.

Spermatophyta One of the divisions of the plant kingdom. The plant body in these is most highly developed and differentiated into root, stem and leaves.

Spermatozoids In algae (Volvox), the female reproductive cell is known as oogonia and the male sex cells are spermatozoids or sperms.

Spermeiogenesis Differentiation of the spermatozoa. Development of sperms.

Spermetocyte The cell that undergoes meiosis to produce four spermatids.

Spices These are certain aromatic and pungent products used for seasoning and flavouring food and various fruit and vegetable preserves.

Spike Here the main axis is elongated and the lower flowers are older, opening earlier than the upper ones, as in raceme, but the flowers are sessile *i.e.* without stalk.

Spinach *Basella* sp. A leafy vegetable.

Spindle Chromosomes move towards the equator of the spindle and arrange themselves in such a way that the cetromere of each chromosomes is attached to the spindle fibre at the equatorial plane.

Spirilization During prophase and before metaphase stage, the chromosomes become shorter and thicker by the process of spirilisation, that transforms the long thin chromatids into thick sort coiled structures in the same way as a thin wire is converted into a coiled spring.

Spirilla These are bacteria with the body spirally wound.

Split plot design An experimental field design in which the land is divided and subdivided into equal blocks corresponding to the number of treatments and subtreatments. The subtreatments are randomized and fixed within each main treatment.

Spongy parenchyma These consist of oval, rounded or more commonly irregular cells, loosely arranged towards the lower epidermis, enclosing numerous, large, intercellular spaces of the vascular bundle.

Spontaneous mutations A mutation appears suddenly, usually as a single individual, and transmits its characters to its progeny as effectively as the normal parent. Mutations may be spontaneous (natural) or induced.

Sporangiophore In fungi *(Mucor)*, the hyphae which are destined to form sporangiophores unbranched and grow upright. The tip of an upright hypha swells to form a spherical sac–like structure called sporangium.

Sporangium In Bryophyta, the sporophyte reproduces asexually by spores. It is very complex structure known as the sporangium.

Spore In Pteridophyta (fern), each sporangium consist of a large slender multicellular stalk and a convex capsule. The capsule is filled with spores. *See* megaspores mother cells.

Spore sac In Bryophyte *(Funaria)*, the columella is surrounded by a cylindrical cavity the spore sac.

Sporogenous In Angiosperms, the young anther shows that in each of its four lobes representing pollen sacs, there is an active merismetic tissue called sporogenous tissue.

Sporogonium in Bryophyta, the oospore grows, divides and redivides to give to a small globular structure called sporogonium.

Sporophyll In Pteridophyta *(Pteridium)*, some of the fronds at maturity bear small rounded uniform or somewhat elongated green to black coloured spots called sori. The frond bearing sori is called sporophyll.

Sporophyte The flowering plants we come across are sporophytes, bearing two kinds of spores *viz.* megaspores and microspores.

Sport A mutation, an abrupt deviation from type.

Sport bud *See* bud mutations.

Spring wood *See* annual rings.

Sprout cells In fungi *(Mucor)*, the mycelium when immersed in a strong sugar solution and is poorly supplied with oxygen becomes into short chain of cells. These cells round off, remain thin walled and are called oidia or sprout cells. This multicellular condition in the form of beaded filaments is called torula condition.

Stalk cell *See* body cell.

Stamen The male reproductive part in the angiosperm also includes Androecium.

Staminate Same as male.

Starch It is an insoluble carbohydrate with chemical formula $(C_6H_{10}O_5)n$. It occurs in all the living tissues of the plants except fungi and is particularly rich in storage organs.

Stele A dicotyledonous stem, consists of pericycle medullary rays, pith and the vascular bundles. In the stem, endodermis is wavy and may or may not be distinct but in roots it forms a distinct water-tight jacket around the stele. The stele is all the tissues inside the endodermis *e.g.* pericycle, vascular bundles, pith and pith rays constitute the stele.

Stem It is the ascending organ of plant and is the direct prolongation of the plumule.

Sterility Failure to produce seed as a result of defective ovules or pollen.

Stigma It is the top portion of the ovary.

Stilt roots A number of roots are produced from the main stem. *See* prop roots.

Stipule These are the lateral appendages of the leaf borne at its base.

Stolon This is a slender lateral breach which originates from the base of the stem, bends down on or into the ground.

Stomata These are very minute openings formed in the epidermis in green serial parts of the plant, particularly the leaves.

Stomatal pits In leaves, the stomata are not spherical but are present as deep pits on the lower surface of leaves.

Stomium In fern, the annulus has an unthickened portion known as the stomium. The capsule wall is thin but it has a specially thickened and cutinized band or ring running round its margin. This ring is the annulus. In Pteridophyta (ferns) below the annulus, the cells of the marginal row are thin walled parenchymatous distended and constitute the stomium.

Stone cells Sometimes here and there in the plant body special types of sclerenchyma may be developed. These are known as the stone or sclerotic cells.

Storage Food is prepared in excess of the immediate need of plants. This surplus food exist in plants in two conditions, either suitable for travelling or suitable for storage. The travelling form is characterised by solubility and the storage form by insolubility in the cell sap.

Storage zone In Bryophyta, it is composed of compact, colourless parenchymatous tissue without any intercellular spaces.

Strain A group of similar individuals within a variety.

Structural gene A cistron.

Struggle for existence When plants or animals are to increase at very high rate, a struggle ensues, resulting in the destruction of a huge numbers of individuals. This kind of competition to live creates a condition in the population where one has to try hard to live under population pressure.

***Strychnos* sp.** This is the nux-vomica plant which yields strychnine from ripe seeds.

Style The middle part of the ovary below the stigma.

Subclimax vegetation A plant community that is prevented from reaching a true climax vegetation. This is caused by localised natural features of the environment of more or less of a permanent nature such as persistently water logged soils or high ground exposed to wind which prevent the growth of climax woodland.

Suberin Is a fatty substance and occurs in the walls of cork cells.

Sucker It is a lateral branch developing from the underground part of the stem.

Sucking roots Same as haustoria.

Summer wood *See* Annual rings.

Sunflower Same as *Helianthus annuus.*

Subsistence agriculture A type of farming in which the output is consumed entirely the farmer and his family leaving only a small proportion for sale.

Subspecies Geographic race; a geographically localised subdivision.

Super females Individuals with three X chromosomes, usually die. The few that survive are so different from other flies that they are easily identified.

Super males Flies with one X, one Y and two sets of autosomes are normal males but flies with one X, one Y and three sets of autosomes are supermales.

Supplementary genes Genes which by themselves have no effect but which qualitatively alter the affect of other gene; modifying genes.

Supplementary ratio In a dihybrid cross, the F_2 phenotypes are seen in the ratio of 9 : 3 : 4.

Suspensor The oospore (angiosperms) divides into two cells: an upper and a lower. The lower one lying towards the micropyle further divides in one direction into a row of cells, called the suspensor.

Suture The junction of two margins of one or more carpels is called the suture.

Swarm spores In Algae, sometimes a large number of zoospores (16–64) are produced in each cell. These are much smaller in size and are known as swarm spores.

Syconus It develops from the hollow, pear shaped, fleshy receptacle which encloses a number of minute male and female flowers.

Symbiosis When two organisms live together, as if they are parts of the same plant, and are of mutual help to each other. They are called symbionts and the relationship between the two is referred to as symbiosis.

Symbols for linked genes While representing linked genes, the two homologous chromosomes are indicated by two horizontal lines with the genes on one chromosomes below the lines *e.g.* $\frac{\text{CS}}{\text{N}}$.

Symmetry A flower is said to be symmetrical when it can be divided into two exactly equal halves by any vertical section passing through the centre. It is also said to be regular or actinomorphic and this conditions is always included in the floral formula.

Symplast The interconnected plasmadesma of the root cells are regarded as a single system, the symplast through which water and minerals move.

Sympodial If in the cymose type, only one lateral branch is produced at a time, the branching is said to be uniperous or monochasial. This type of branching is otherwise called sympodial.

Synapsis The pairing of homologous chromosomes of maternal and paternal origin in the prophase of meiosis.

Syncology It is the study of plant community with relation to the environment. Some workers call it as plant sociology.

Synergids These successive mitotic divisions of the embryo sac nucleus give rise to eight haploid nuclei within the embryo sac. One of the three nuclei at the microphyle end of the sac becomes the egg or the female gamete and the other two become synergids. Also *See* egg apparatus.

Syngamy The union of sex cells (gametes) in the production.

Syngenesious stamens When the anthers are united into one bundle or tube, but the filaments are free.

Synteny Occurrence of two loci on the same chromosome independent of distance that separates them suitably.

Synthetic chimeras Same as graft chimeras.

Synthetic variety A variety produced by crossing inter se a number of genotypes selected for good combining ability in all possible hybrid combination with subsequent maintenance of the variety by open pollination.

Synthetic variety seed Synthetic varieties yield more than inbred lines or the open pollinated varieties but do not yield as the sing cross or double cross hybrids but they have the advantage of not showing any reduction in yield for nearly three generations. It is therefore practicable that sufficient seeds are distributed every year to farmers once in every four years whereas single cross or double cross hybrid seeds have to be distributed every year.

Syzygium aromaticum Clove. Medicinal use.

Tagetes patula Fam. Compositeae. Marigold. Ornamental flower.

Talaeopora In this moth (insect), the females have 59 chromosomes and the males have 60 chromosomes in their somatic cells.

Tamarind Fam. Caesalpieae. *Tamarandius indica.* Fruits widely used for sour preparations. A shade tree.

Tandem duplication A condition in which the duplicated segment is directly adjacent to the normal region in the chromosomes.

Tannis These are a group of complex compounds widely distributed in plants. They commonly occur in single isolated cells or in small groups of cells in almost all parts of the plant body.

Tapetum In Angiosperms flower, the sporagenous tissue is surrounded by a layer of nutritive cells called tapetum.

Tapioca Is a large fleshy root of *Manihot esculentus,* is a perennial shrub. It is used as food in the form of flour.

Tap root The radical forms the primary root. If it persists and continues to grow, as in dicotyledons, giving rise to the main root of the plant, it is the tap root.

Taxis In induced movement of locomotion, if the taxic movement the stimuli is a chemical, it is a chemotaxis and it influenced by light then it is called phototaxic.

Taxonomy It deals with the description, identification and naming of plants and their classification into different groups according to their resemblances and differences mainly in their morphological characteristics. This is also called the systematic botany. It deals with the nomenclature description and classification of plants and animals.

Tea Fam. Cammilanceae. *Thea (Camellia) asamica* and *T. sinensis*. The dried and processed leaves of these species and many hybrids are used as common beverage.

Teak *Tectona grandis*. A very valuable timber tree.

Tegmen The two integuments of the ovule develop into two seed-coats, of which the outer is called the testa and the inner is the tegmen.

Telophase The last stage in cell division before the nucleus returns to the resting phase.

Teminism A phenomenon in which DNA is synthesised on RNA template discovered in cercoma virus by Tenin and Baltimore.

Template A model mold or partner; DNA acts as a template for RNA synthesis.

Tendril These are slender, leafless, spirally coiled structures help the plant to climb.

Tentacles In sundew *(Drocera* sp.*)*, each leaf is covered on the upper surface with numerous glandular hairs known as tentacles.

Terminal chiasmata Chiasmata found at the ends of the chromosomes, they are said to be in terminal position.

Terminalization As contraction of chromosomes proceeds, the chiasmata tend to loose their original position and move towards the end of the chromosomes. This movement of chiasmata is terminalization. Expansion of the association of the two pairs of chromatids one side of the chaismata at the expense of that on the other side.

Testa *See* tegmen.

Test cross A cross between a heterozygote and a recessive homozygote to test for homozygosity or for linkage.

Testing inbreds for general combining ability Those inbred lines with good general combining ability are crossed in all possible combinations and the specific ability of a line to combine well with another line is determined from the yield of the progenies available.

Tetrad At telophase II of meiosis, the four groups of chromosomes now organised as four nuclei. The four nuclei are collectively known as tetrad. Each of them contains a haploid set of chromosomes.

Tetradynamous In such flowers, there are six stamens of which four are long and two short.

Tetraploid Any organism having four sets of basic number of chromosomes (4x).

Tetrarch In a dicot root, there are two to five xylem and an equal number of phloem bundles. A root with four bundles is called tetrarch.

Tetrasomic An organism having two extra chromosomes other than to their normal somatic complement (2n+2).

Thalamus The swollen end of the axis of flower with the floral leaves inserted on it. A typical flower of Angiosperms has a short or long stalk which is flattened or conical at the tip and is called thalamus.

Thallophyta In this group, the plant body is most simple. It is either unicellular or multicellular, when multicellular it is not differentiated into organs of different types of tissues.

Theca In Bryophyta, it is the body containing spores and is a part of the capsule and shows much internal differentiation.

Theory of natural selection According to this theory, many more individuals of a species are born than can possibly survive and consequently there is always a struggle for existence. If hereditary differences occur within the wild (original) species of plants, nature will eliminate some and select others.

Thermal neutrons Fast neutrons can be converted into very slow neutrons called thermal neutrons. This can be done by reducing their velocity to that of thermal energies.

***Thespesia* sp.** Fam. Malvaceae. Portia tree, a shady tree.

Thevetia peruvian a Fam. Apocynaceae. Yellow oleander. Medicinal and ornamental.

Thigmotropism Same as haptotropism.

Thorn apple *Datura* sp.

Three bean experiment That all the conditions of proper light, air and temperature and moisture are necessary for germination of seed.

Thyme Fam. Labiateae. *Thymus* sp. Source of thyme oil and thymol.

Thymine A constituent of DNA ($C_6H_6N_2O_2$). A pyrimidine base of the deoxynucleic acid, first isolated from the thymus.

Tissue It is a group of cells of the same type or of the mixed type, having a common origin and performing an identical function.

Tonoplast The vacuole contains water, solutions of salts, organic molecules and waste products of the cell's metabolism. It is surrounded by a differentially permeable membrane called tonoplast.

Top cross The inbred variety cross made for the preliminary testing of a large number of inbreds.

Topoisomerase (gyrase) Enzyme capable of making cut in the backbone of one strand of the DNA, double helix during replication and rapidly repairing such cuts after unwinding.

Totipotent cell An undifferentiated living cell which when suitably transplanted can develop into a complete embryo.

Trabeculae In the theca (Bryophyta), the spore sac is followed on the outside by an air space which is traversed by multicelluar filaments called trabeculae.

Trachae Also called vessels. These are not single celled structures (in contrast to trachieds) but are composed of elongated cylindrical cells or vessel elements which are arranged end to end in longitudinal rows and from which the end walls have been digested either completely or partially.

Trachieds These are elongated tubes like cells with large lumen and tapering chisel like cells.

Transfection The uptake of DNA by eukaryotic cell followed by the incorporation of genetic markers present in this DNA into cell's gemon.

Transfer of recessive gene The two varieties concerned are crossed. Suppose the F_1 is susceptible to rust. It is backcrossed with the susceptible variety. The progeny from the backcross consists of two genotypes, SS and Ss in equal numbers. Since all the plants are susceptible, the heterozygotes with recessive allele for resistance cannot be picked for further backcrossing. All the plants are therefore selfed and the resistant plants are picked out for the second backcross with the susceptible variety. The progeny from the second backcross consists entirely of susceptible plants. These are heterozygous with recessive alleles for resistance and they are therefore backcrossed with the susceptible variety of the third time. The progeny for the third backcross, like that for the first backcross, consists of two genotypes, SS and Ss in equal numbers. The cycle is repeated till the progenies are like the susceptible variety. After the required number of backcross, all the plants are selfed and the plants which are resistant to rust are selected.

Transfer RNA Amino acid specific RNA which transfers activated amino acids to RNA where protein synthesis takes place.

Transgressive factors The appearance in the F_2 of later generation of individuals, showing a more extreme development of the trait than shown in either original parent.

Transgressive segregate In hybridization, occasionally the recombination of genetic factors leads to the production of new and desirable characters not found in either parent. For example, the progenies may be taller than the taller parent or earlier than the earlier maturing parent. Such transgressive segregates may enable the breeder to attain his objective completely.

Trans-heterozygote A cell or an organism with two mutations in the trans configuration (one on each chromosome).

Translocation Food materials are mostly prepared in the leaves. From there they are translocated to the storage organs which often lie at a considerable distance for this purpose, there are definite and distinct channels extending through the whole length of the plant body. Simple translocation is chromosomal aberration which involves the transfer of the end of one chromosome to the end of another homologous chromosome.

Transpiration It is giving off of water vapour from the internal tissues of living plants through the aerial part such as the leaves, green shoot etc., under the influence of sunlight regulated by the protoplasm. Water is taken by plants in large quantities through the root system. The water loss from the living plants by evaporation is called transpiration.

Trichomes Usually the epidermis develops multicellular branched or unbranched hairs trichomes.

Trichosanthes anguina Fam. Cucurbitaceae. Snake gourd. A vegetable.

Trifolium Gametophytic incompatibility is prevalent in this genus.

Trihybrid An individual heterozygous for three pairs of alleles.

Trihybrid ratio The 27 genotypes produced by a trihybrid cross have eight visibly different types (phenotypes).

Triploid An individual having three sets of basic number of chromosomes in its somatic complement.

Tristichous It is phyllotaxy 1/3. The fourth leaf stands vertically over the first one and the genetic spiral makes one turn to reach that leaf and it involves three leaves.

Trisomic An organism having one chromosome of the diploid complement in triplicate, hence having 2n+1.

Tristyly In Lythrus, the styles are of three lengths, short, medium and long; and the filaments are of three lengths corresponding to the lengths of the styles. Anyone plant has one style length and four stamens with two different lengths of the filaments. Pollinations are compatible only between stigmas and anthers at the same level. Thus each type of plant can effectively fertilize the other two types.

Triticale An example of indiced polyploid.

Triticum aestivum The hexaploid common bread wheat.

Trivalent An association of three homologous chromosomes held together by chiasmata during the period from diplonema to first metasphase.

Tropisms Tropic movements are the movements of plant organs influenced by external stimuli, particularly light, contact, gravity and moisture.

Tropophyte A plant that changes its physiology to suit changing climatic conditions. Such plants develop hydrophytic tendencies when plenty of water is available to them but adopt xerophytic properties such as loosing leaves during seasonal draught.

Tube cell The nucleus of the pollen grain divides into two–a large and a small nucleus. The larger nucleus lies in the main cytoplasm of the pollen grain and forms the large vegetative or tube cell.

Tube nucleus *See* generative nucleus.

Tuber A underground modified stem.

Tuberculated rhizoids *See* smooth rhizoids. While others have distinct peg-like small outgrowths or tubercles on the inner surface of the wall and are known as tuberculated rhizoids.

Tuberose Fam. Amaryllidaceae. *Polianthes tuberosa.*

Turgid This is the pressure exerted on the liquid contents of the cortical cells of the root, under fully turgid condition, forcing a quantity of them into the xylem vessels and through them upwards into the stem upto a certain height.

Turgor pressure The outward pressure exerted by the cell wall by the fluid contents of the cell.

Turmeric It has a modified underground stem known as rhizome.

Twiners These structures bodily twine round some support and are the climbing aids to the plants.

Tyloses These are protusions of parenchyma cells which enter the lumen of the vessels and trachieds through the pits in their walls.

Ulothrix A green filamentous algae.

Ultraviolet The only non-ionizing radiation capable of producing mutants.

Umbel In this inflorescence, the primary axis is shortened and it bears at its tip a ground of flowers which have pedicels of more or equal lengths so that the flowers are seen to spread out from a common point.

Umbellifereae A family of angiosperms.

Unit membrane A membrane formed of two layers of lipid molecules sandwiched between the two layers of protein molecules. It forms the outer boundary of cell and almost all cellular organneles.

Univalent An unpaired body at the first meiotic division corresponding to a single chromosome in the complement.

Uracil A pyrimidine base occurring in RNA ($C_4H_4N_2O_2$).

Urena lobata Fam. Malvaceae.

Use of hybrid vigour in breeding It necessitates the production of hybrid seeds in such large quantities that the F_1 progeny can be grown on a field scale.

Vacuoles Cytoplasmic structures for storage of excess water, waste products, soluble pigments etc.

Valvate When the members of a whorl are in contact with each other by their margins, or which they lie very close to each other, but do not overlap.

Variegation With inheritance studies in *Mirabilis jalapa,* it is observed that this is a hereditary character, determined by stable, self-duplicating extra nuclear particles called plastids.

Variations Differences among individuals of a population due to differences in their genetic composition or the environment in which they grow.

Variety A subdivision of a species. An agricultural variety is a group of individuals within a species which are distinct in appearance or performance from other varieties within the same species.

Vascular bundle A vascular bundle of a dicotyledonous plant stem. When fully formed, consists of three wall defined tissues; xylem or wood, phloem or bast and cambium. They have different kinds of tissue elements.

Vavilov A Russian Scientist from the study of variations in crop plants and their wild allies, found that there are some regions which are the centres of genetic diversity. These centres, he postulated, were the centres of origin of a species, which he called the primary centres for origin of species.

Vegetative cells The somatic cells.

Vegetative reproduction Asexual reproduction by buds, grafts, cuttings, layering, etc.

Velamen The hanging root has a covering of special tissue, usually 4 or 5 layers, in thickness called velamen.

Venation The arrangements of the veins and the veinlets in the leaf blade.

Ventre In *Marchantia,* the archegonium is a flash-shaped body consisting of a swollen basal portion, the venter and a narrow tubular portion, the next.

Vernalization Some winter plants like wheat are subjected to low temperatures treatment during their germination stages to make them flower early in the following summer. This method of pretreating of partially germinated grains to low temperatures is called vernalization.

Versatile When the filament is attached to the back of anther, the attachment is versatile.

Verticillaster This is a special form of cymose inflorescence. In it here is a cluster of sessile or almost sessile flowers in the axil of a leaf forming a false whorl at the node.

Vessels These are not single celled structures in contrast to the trachids but are composed of elongated cylindrical cells or vessel elements, which are arranged end to end in longitudinal rows and from which the end walls have been digested either completely or partially. These are also called trachae.

Vexillary In this type of aestivation, there are five petals of which the posterior ones is the largest and it almost covers the two lateral petals, and the latter in their turn nearly overlap the two anterior or smallest petals.

Vicinism Out-crossing; natural cross-pollination.

Vigna sinensis Fam. Papilionaceae. Cow pea. A pulse crop.

Village seed it is the seed produced on a large scale in a village by certified seed growers under the general supervision of departmental staff. It maintains genetic identify and purity and can be distributed to formers.

Vitastic theory It is believed by some scientists that the activity of living cells *e.g.* wood parenchyma and medullary cells surrounding xylem is responsible for the rise of sap through the plant body.

Vitamin Organic compounds found in small amounts in all living organisms and which are essentially required by body for its normal working.

Vivipary The seed germinates inside the fruit while it still on the parent tree and is nourished by it.

Volvox It is an alga most complex but beautiful and fascinating of the motile colonial algae. This genus consists about twenty species distributed throughout the world in fresh ponds and lakes.

V-shaped chromosome If the centromere is located at the middle of a chromosome, it is called V-shaped.

Vulgare It is the hexaploid group of Triticum having 2n+42.

Wall pressure The inward pressure exerted on the cell contents by the stretched cell wall is called wall pressure. The turgor pressure is put by the cell contents on its wall while the wall pressure is put on the contents by the wall.

Water relations Water forms 80 to 90 per cent of fresh weight of plant. It can exist as (a) true solution in vacuoles (b) dispersal medium for colloidal system and (c) water held in the pores of the cell wall.

W-chromosome *See* Z chromosome.

Wood *See* xylem.

Wood fibres Also called xylem fibres. These are ordinary schlerenchyma fibres occurring in association with other wood elements. They are elongated narrow cells with heavily thickened walls and are pointed at both ends.

Wood parenchyma or parenchyma Parenchymatous cells found in association with xylem elements. Like other parenchymatous cells, they are thin-walled and living.

Wood rays The vascular rays or secondary medullary rays which lie in the xylem are called xylem or wood rays.

X The basic number of chromosomes in a polyploid.

X_1, X_2, X_3 First, second, third etc. generation following irradiation with X-rays.

Xanthophyll The colouring matters (pigments) associated with chloroplasts are xanthophyll (yellow) and carotene (orange red) which occur in different proportions.

X-chromosome The sex chromosome which is associated with an identical homologue in the female (XX) and with the dissimilar partner in the male (XY or O with no partner in the male X).

Xenia Effect of pollen on embryo and endosperm.

Xerophyte A plant that can survive scarcity of water or draught. Xerophytes are typically found in desert soils and semiarid climates and include cacti. They show adaptations to such habitats including long or enlarged roots, water storage organs and leaves that are thickened and waxy coated reduced to narrow spines or deciduous during dry seasons. These plants are typically stout stemmed, thorny or succulent.

X-rays The mutagenic action of X-rays was first demonstrated by Muller on Drosophila and subsequently by Stdler in 1928 on barley and maize. Since then it has now been recognised as an important agent for induction of mutations in plant breeding.

XX-XO type The XX-XO types of sex chromosomes are found in Protenor bug.

Xx-xy type These types of sex chromosomes are found in Drosophila and other animals including man.

Xylem Also called wood. This lies toward the centre and is composed of four main types of cells (*i*) tracheids (*ii*) vessels or tracheae (*iii*) xylem fibres or wood fibres and (*iv*) xylem parenchyma or wood parenchyma.

Y-chromosome In male Drosophila, there is only one rod shaped chromosome, the member of this pair (sex-pair) is J-shaped *i.e.* hook shaped, or like a rod with a bent end, this is the V-chromosome.

Z-chromosome The sex chromosome which is associated with an identical homologue in the male (ZZ) and with a dissimilar partner in the female (ZW) or no partner in the female (Z).

Zoophilly In such act of pollination, these animals play the role of fertilizing the flowers by carrying the pollen from one flower to another.

Zoosporangium The cell which forms the zoospores in Algae is the zoosporangium.

Zoospores In *chlamydomons* (alga), the asexual reproduction takes by the zoospores.

Zo-Zz type This type of sex chromosomes are found in birds including the domestic fowls. The female has an unlike pair of chromosome, ZW and form eggs of two sorts, one with a W and another with Z chromosome.

Zygomorphic When the flower can be divided into two similar halves by one such vertical axis only, it is said to have zygomorphic symmetry.

Zygonema Also called zygotene, during this stage of prophase (meiosis) the pairs of chromosomes begin to come together by what appears to be a force of mutual attraction. The pairing of homologous chromosomes is called synapsis, and it begins at one or more points and gradually proceeds along their whole length in a zipper-like fashion.

Zygosporangium *See* Plant Pathology.

Zygospores *See* conjugation.

Zygote This consists of in the fusion of two reproductive units called gametes. Gametes are always unicellular and microscopic in size. Two gametes of opposite sexes tuse together and the product of such a fusion is a new cell called the zygote, which develops into a new plant.

Zygotic mutation A mutation occurring in the zygote shortly after fertilization.

Zygotic number Same as somatic number.

Zymogen These are very minute granules secreted by the enzymes.

ZW-ZZ type In this case, the female is the heterogametic sex and the male is the homogametic one *e.g.* in case of the moth *Taleaoporia.*